PREFACE

➢ The present book is specifically designed for the students who prepare for various competitive exams like **engineering and medical entrances.**

➢ The primary objective of this book is to provide fundamental knowledge of the chemical elements.

➢ In recent years, the idea about chemical elements has been considered extremely important for students desirous of pursuing chemical sciences courses.

➢ This book is designed to expand on the foundation that has been acquired in understanding the fundamental aspects of chemical elements and especially to enable one to understand the physicochemical properties of elements.

➢ Moreover, the subject matter has been arranged systematically in a simple language.

CONTENTS

PERIODIC PROPERTIES

1. Periodic table is a table of elements in which the elements with similar properties are placed together.

2. Dobereiner's law of triads was the classification of elements into groups of three elements each with similar properties such that the atomic weight of the middle element was the arithmetic mean of other two.

    Eg: Ca, Sr, Ba;   Cl, Br, I, etc.

3. Newland's law of octaves was an arrangement of element in order of increase in atomic weight.

4. Mendeleev's periodic table is based upon Mendeleev's periodic law which states that the physical and chemical properties of the elements are a periodic function of their atomic weights.

5. Mendeleev's original periodic table consists of 8 vertical columns called groups I-VIII & zero and seven horizontal rows called periods 1-7.

6. Elements of group IA are called alkali metals while those of groups IB are called coinage metals (Cu, Ag, Ao).

7. Group VIII consists of 9 elements in form of three triads placed in periods 4, 5 and 6.

8. Moseley measured the frequencies of X-rays emitted by a metal when bombarded with high speed electrons. He concluded that atomic number was a more fundamental property of an element than its atomic weight. He forms the modern periodic law.

9. Modern periodic law states that "the physical and chemical properties of the elements are a periodic function of their atomic number".

10. Magic Numbers: When the elements are arranged in order of increasing atomic number, it was observed that the properties of elements were repeated after certain regular intervals of 2, 8, 8, 18, 18 and 32. These numbers are

3

called magic numbers and cause of periodicity in properties is due to repetition of similar electronic configuration at certain regular intervals of 2, 8, 18, 18 and 32.

11. Structural features of the long form of the periodic table.

(i)   It consists of 18 vertical columns called groups and 7 horizontal columns called periods.

(ii)   Elements of groups 1, 2, 13 to 17 are called representative elements.

(iii)   Elements of groups 3-12 are called transition elements.

(iv)   The 14 elements with atomic number (z) 58 to 71 are called lanthanides (or) rare earth elements and are placed at the bottom of the periodic table.

(v)   The 14 elements Z=90 to 103 are called actinides are placed at the bottom of the periodic table.

(vi)   The eleven elements with Z=93 to 103, which occur in the periodic table after uranium and have been prepared by artificially. These are called transuranics. These are all radioactive elements.

(vii)   The elements belonging to a particular group are said to constitute chemical family which is usually named after the name of the first element.

Eg: Boron family (group 13)

Carbon family (group 14)

Nitrogen family (group 15)

Oxygen family (group 16)

In addition to this, some groups have typical names:

Eg: Elements of group 1 are called alkali metals

Elements of group 2 are called alkaline earth metals

Elements of group 16 are called chalcogens

Elements of group 17 are called halogens

Elements of group 18 are called noble gases (or) zero group

The long form of periodic table contains 7 periods. These are

1st period ($_1$H-$_2$He) shortest period – contains only two elements.

2nd period ($_3$Li-$_{10}$Ne) and third period ($_{11}$Na-$_{18}$Ar) contains 8 elements each and are called short periods.

4th period ($_{19}$K-$_{36}$Kr) and 5th period ($_{37}$Rb-$_{54}$Xe) contains 18 elements each and are called long periods.

6th period ($_{55}$Cs-$_{86}$Rn) contains 32 elements and is the longest period.

7th period ($_{87}$Fr-) is however incomplete and contains at present only 25 elements.

In another way, the periodic table classified into four blocks. They are:

1. s-block

2. p-block

3. d-block

4. f-block

1) **s-block:** Elements of groups 1 and 2 in which the last electron enters the s-orbital of the valence shell are called s-block elements.

2) **p-block:** In which the last electron enters p-orbital of the valence shell are called p-block elements. Elements of groups 13-18 i.e. IIIA to 'O' group.

3) **d-block:** Elements of groups 3-12 in which the last electron enters the d-orbital of the penultimate shell are called d-block elements (or) transition elements. There are 3 complete series and one in complete series of d-block elements. These are:

5

3d series which contains ten elements with Z=21-30 ($_{21}$Sc-$_{30}$Zn)

4d series which contain ten elements with Z=39-48 ($_{39}$Y-$_{48}$Cd)

5d series which contain ten elements with Z=57 and 72-80 ($_{57}$La, $_{72}$Hf – $_{80}$Hg)

6d series which is incomplete at present and contains only nine elements.

4) **f-block:** In these elements, the f-orbital of the anti-penultimate shell is being progressively filled up. These are called f-block elements (or) inner transition elements. There are two series of f-block elements. They are:

Lanthanides which consists 14 elements with Z=58-71 ($_{58}$Celectron$_{71}$Lu).

Actinides which consists 14 elements with Z-90-103 ($_{90}$Th-$_{103}$Lr).

12. **Diagonal Relationship:** Certain elements of 2nd period show similarity with their diagonal elements in the 3rd period as shown below:

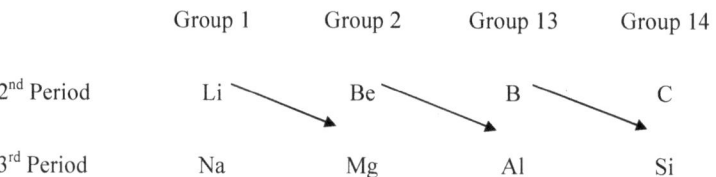

	Group 1	Group 2	Group 13	Group 14
2nd Period	Li	Be	B	C
3rd Period	Na	Mg	Al	Si

Thus, Li resembles Mg, Be resembles Al and B resembles Si. This is called diagonal relationship and is due to the reason that these pairs of elements have almost identical ionic radii and polarizing power.

13. **Anamalous behavior of the first element of a group:** The first element of a group differs considerably from the rest of the elements. This is due to :

(i)   Small size

(ii)   High electro-negativity

(iii)   Non-availability of d-orbitals for bonding.

14. **Periodic properties:** Properties which are directly or indirectly related to the electronic configuration of the elements and show a regular gradation when

we move from left to right in a period or from top to bottom in a group are called periodic properties. Some of periodic properties are atomic size, ionization energy, electron affinity, electronegativity, valency, density, atomic volume, melting and boiling points etc.

a) Atomic Size: It refers to the distance between the centre of the nucleus of the atom to the outer most shell containing electrons.

b) Ionization energy (IE): IE is the amount of energy required to remove the most loosely bound electron from an isolated gaseous atom.

$$M(g) + 1E \quad \rightarrow \quad M^+ (g) + electron$$

The amount of energies required to remove the 1st, 2nd, 3rd etc. electrons from the isolated gaseous atom are called successive IE and are designated $1E_1$, $1E_2$, $1E_3$ etc.

$$IE_3 > 1E_2 > IE_1$$

The factors affecting the ionization energies:

(1) Nuclear charges (2) Atomic size (3) Penetration effect of the electrons (4) Effect of exactly half filled and completely filled orbitals.

c) Electron Affinity (EA): EA is the amount of energy released when a neutral isolated gaseous atom accepts an electron to form gaseous anion.

$$X(g) + e \quad \rightarrow \quad X^- (g) + EA$$

EA depends upon (i) Atomic size (ii) Nuclear charge, (iii) Electronic configuration.

d) Electronegativity (EN): EN is the tendency of an atom in a molecule to attract the shared pair of electrons towards itself. It depends upon (i) atomic size (ii) nuclear charge.

e) Valency: of an element is the number of electrons gained (or) lost (or) shared with other atoms in the formation of compound.

7

The valency of representative elements is equal to the number of electrons in the outermost shell.

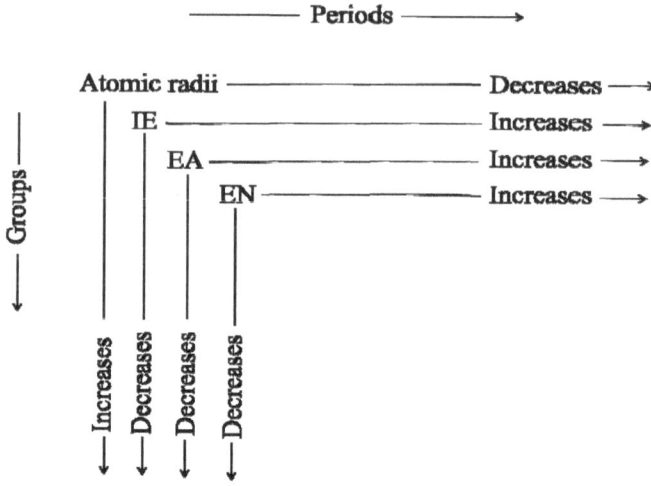

- An atomic and ionic radius increases from top to bottom due to addition of new shell at each succeeding element of representative elements.

- Within a period, the increase in IE is not regular.

  Eg: 2nd period, as we go from Li → Be, the 1E, increase partly due to increased nuclear charge. From Be → B, the 1E decreases because in case of B, a less tightly held p-electron is to be removed while in Be, a more tightly held s-electron is to be remove.

- Within a group EA decreases from top to bottom due to increasing atomic size.

- Within a period EA increase from left to right due to increasing nuclear charges.

- Within a group EN decreases from top to bottom due to increasing atomic size.

- Within a period, EN decreases from left to right due to increasing nuclear charge.

- For the same element, the EN depends upon the state of hybridization. For example, EN of C in the three states of hybridization varies as $SP > SP^2 > SP^3$. As the S-character of hybrid orbital's decreases, the EN also decrease.

ALKALI METALS

The elements

Lithium	-	Li (Z=3)
Sodium	-	Na (11)
Potassium	-	K (19)
Rubidium	-	Rb (37)
Cesium	-	Cs (55)
Franium	-	Fr (87)

IA group elements are collectively called alkali metals since their oxides and hydroxides from strong alkalies.

Eg: NaOH, KOH etc.

Physical Properties:

1) Electronic configuration: nS^1 (or) $(n-1) S^2 p^6 nS^1$

   Alkali metals are S-block elements. These contain only one electron in the s-orbital.

2) Density of alkali metals is quite low and increase group from Li to Cs due to an increase in their atomic mass. K, however, is lighter than Na which is due to abnormal increase in its atomic size.

   $$Li < Na > K < Rb < Cs < Fr$$

3) Melting point (M.P) and Boiling Point (B.P) of alkali metals decrease with increase in atomic number due to weakening of metallic bond. Fr is liquid at room temperature.

   $$Li > Na > K > Rb > Cs > Fr$$

4) Softness: Alkali metals are soft, malleable and ductile solids which can be cut with knife.

5) Atomic volume of alkali metals decrease from Cs to Li

   $$Li < Na < K < Rb < Cs$$

6) Ionization energy (IE) decrease down the group from Li to Fr

$$Li > Na > K > Rb > Cs > Fr$$

7) Electropositive Character (EP): Alkali metals are strong electropositive because of their low IE.

8) Atomic radii of alkali metals increase from Li to Cs

$$Li < Na < K < Rb < Cs$$

9) Crystal structure: All alkali metals posses' body centered cubic (BCC) structure with co-ordination no. 8.

10) Flame Colouration:

Li	-	Crimson
Na	-	Golden yellow
K	-	Pale violet
Rb	-	Purple
Cs	-	Violet

11) Photoelectric effect: K and Cs show photoelectric effect and are used in photoelectric cells.

12) Electrical conductivity: The metals are good conductors of heat and electricity. Electrical conductivity increase from Li^+ to Cs^+.

$$Li^+ < Na^+ < K^+ < Rb^+ < Cs^+$$

13) Reducing Character: The metals are good reducing agents, follows the order:

$$Na < K < Rb < Cs < Li$$

14) Mobility of ions: The metal ions exists as hydrated ions in the aqueous solution. The degree of hydration, decrease with ionic size as we move from Li^+ to Cs^+. The mobility of ions inversely proportional to size of their hydrated ions, Li has the lowest ionic mobility.

15) Chemical Properties: Due to low IE, the metals are chemically very reactive and hence do not occur free in nature.

  1) Reaction with Water: All metals are readily react with H_2O evolving H_2 gas.

$$2M + 2H_2O \rightarrow 2MOH + H_2$$

The reactivity increases from Li to Cs.

$$Li < Na < K < Rb < Cs$$

2) Hydroxides: The alkalihydroxides are strong bases and their basic strength increases from LiOH to CsOH.

$$LiOH < NaOH < KOH < RbOH < CsOH$$

3) Reaction with Oxygen: Alkali metals react with O_2 from different oxides.

Eg:	LiO_2	-	Lithiom oxide
	Na_2O_2	-	Sodium peroxide
	KO_2	-	Potassium superoxide
	RbO_2	-	Rubidium superoxide
	CsO_2	-	Cesium superoxide

4) Reaction with hydrogen: All alkali metals when heated with H_2 form hydrides.

$$2M + H_2 \rightarrow 2M^+ H^-$$

The reactivity of hydrides decrease from Li to Cs.

$$Li > Na > K > Rb > Cs$$

5) Reaction with halogens: All alkali metals react with halogens to form their halides.

$$2 M + X_2 \xrightarrow{\Delta} 2M^+ X^-$$

The reactivity order of alkali metals $Li < Na < K < Rb < Cs$

6) M.P: M.P decreases in the order $F_2 > Cl_2 > Br_2 > I_2$. Since their lattice energies decrease as the size of the halide ion increases.

$$NaF > NaCl > NaBr > NaI$$

7) Reaction with N_2: Only Li reacts with N_2 to form lithium nitride:

$$6 Li + N_2 \xrightarrow{\Delta} 2Li_3N.$$

8) Reaction with Sulphide & Phosphorus (P): Alkali metals when heated with S_8 & P_4 form their respective sulphides and phosphides.

$$16 \, M + S_8 \xrightarrow{\Delta} 8 \, M_2S$$

$$12 \, M + P_4 \rightarrow 4 \, M_3P$$

9) Solubility in liquid ammonia: The metals react (or) dissolve in liquid NH_3 to give deep blue solutions due to the presence of solvated electrons in the solution.

$$M + (X+4) \, NH_3 \rightarrow M^+ (NH_3)_x + \text{electron } (NH_3)y.$$

10) Nature of the carbonates and bicarbonates: Li_2CO_3 is much less stable and decomposes on heating to give LiO_2 & CO_2.

$$Li_2CO_3 \xrightarrow{redheat} Li_2O + CO_2$$

Carbonates stability increases from $Na_2CO_3 + Cs_2CO_3$.

$$Li_2CO_3 < Na_2CO_3 < K_2CO_3 < Rb_2CO_3 < CsCO_3$$

Similarly, the bicarbonates stability increases from Li to Cs.

$$LiHCO_3 < NaHCO_3 < KHCO_3 < RbHCO_3 < CsHCO_3$$

11) Nature of nitrates: $LiNO_2$ on heating decomposes to give NO_2 and O_2. All nitrites are soluble in water.

$$4 \, LiNO_3 \xrightarrow{\Delta} 2 \, Li_2O + 4NO_2 + O_2$$

$$2 \, NaNO_3 \xrightarrow{\Delta} 2 \, NaNO_2 + O_2$$

12) Nature of Sulphates: Li_2SO_4 is insoluble while the sulphates of other alkali metals are soluble in water.

- Li shows diagonal relationship with Mg since they have almost the same polarizine power.
- Li can't be stored in kerosene oil since it floats over the surface due to its very low density. Li is usually kept in paraffin wax.
- All alkali metals dissolve in Hg to form amalgams.
- A mixture of Na_2O_2 and dil. HCl is called oxone and is used for bleaching delicate fibres.

ALKALINE EARTH METALS

The elements

Be	-	Beryllium (z=4)
Mg	-	Magnesium (12)
Ca	-	Calcium (20)
Sr	-	Strontium (38)
Ba	-	Barium (56)
Ra	-	Radium (88)

These are commonly called alkaline earth metals because metals oxides are alkaline in nature and are found in earth's crust.

Physical Properties

1) Electronic configuration. ns^2: They contains 2 electrons in s-orbital.
2) Metallic properties: They are silvery white metals, soft in nature. All of them have high electrical and thermal conductivities.
3) Density: They are denser than alkali metals. The density first decrease from Be to Ca and then steadily increase from Ca to Ra.
4) M.P and B.P.: They have highest M.P & B.P.
5) Atomic and ionic radii of alkaline earth metals are fairly large, due to higher nuclear charge.
6) Atomic volume: Atomic volume of alkaline earth metals increases from Be to Ra as atomic radius increases.
7) Ionization energy (IE): The IE_1 and IE_2 of alkaline earth metals are fairly low. The IE_2 is almost double of the IE_1. IE decrease from Be to Ra due to increase in atomic size:

$$Be > Mg > Ca > Sr > Ba > Ra$$

8) Oxidation state: The alkaline earth metal has +2 oxidation state.
9) Electropositive Character (EP): Alkaline earth metals are strongly EP since they have a strong tendency to lose both the valence electrons.

10) Flame colouration:

Ca	-	Brick red
Sr	-	Crimson red
Ba	-	Apple green
Ra	-	Crimson

Be and Mg don't have any characteristic colours because they have high IE.

Chemical properties:

Due to low IE, alkaline earth metals are fairly reactive. However, their chemical reactivity is lower than those of alkali metals.

1) Reaction with water: They react with H_2O evolving H_2 gas.

$$M + 2H_2O \rightarrow 2\,M\,(OH)_2 + H_2$$

Chemical reactivity of the metal with H_2O increases from Mg to Be.

2) Reaction with acids: Alkaline earth metals except Be, displace hydrogen from acids.

$$M + H_2SO_4 \rightarrow MSO_4 + H_2$$

Reactivity increases from Mg to Ba.

3) Reaction with Oxygen: The affinity for oxygen increases from Be to Ba.

$$2M + O_2 \overset{\Delta}{\rightarrow} 2MO \ (M = Be, Mg \ (or) \ Ca)$$
$$(\text{Metaloxide})$$

$$M + O_2 \overset{\Delta}{\rightarrow} MO_2 \ (M = Ba, Sr \ (or) \ Ra)$$
$$(\text{Metal peroxide})$$

4) Reaction with hydrogen: All the alkaline earth metals except Be, combine with H_2 directly on heating to form metal hydrides.

$$M + H_2 \overset{\Delta}{\rightarrow} MH_2$$

BeH_2 can be prepared by reducing $BeCl_2$ with $LiAlH_4$.

$$2\,BeCl_2 + LiAlH_4 \rightarrow 2\,BeH_2 + LiCl + AlCl_3$$

All hydrides react with water to evolve H_2 and thus behave as strong reducing agents.

5) Reaction with N_2: When heated with N_2 form their respective nitrides.

$$3\,M + N_2 \rightarrow M_3N_2$$

The ease of formation of nitrides increases from Be to Ba.

6) Reaction with carbon: When heated with carbon, alkaline earth metals form their respective carbides.

$$M + 2\,C \xrightarrow{\Delta} MC_2$$

7) Reaction with halogens: When heated with halogens, all the alkaline earth metals form halides.

$$M + X_2 \rightarrow MX_2$$

8) Reducing character: All the alkaline earth metals are strong reducing agents. Reducing characters increases from Be to Ra.

9) Basic strength of oxides and hydroxides: Basic strength increases from Be to Ba.

10) Solubility of hydroxides: The solubility of the hydroxides of alkaline earth metals increases from Be to Ba.

11) Solubility of sulphates: The solubility of sulphates of alkaline earth metals decrease from Be to Ba.

12) Solubilities of bicarbonates and carbonates: On heating, bicarbonates forming carbonates with the evolution of CO_2.

$$M\,(HCO_3)_2 \xrightarrow{\Delta} MCO_3 + CO_2 + H_2O.$$

The solubility of carbonates decreases form Be to Ba.

13) Thermal stability of carbonates: The carbonates of alkaline earth metals decompose on heating forming metal oxide and carbondioxide.

$$MCO_3 \xrightarrow{\Delta} MO + CO_2$$

14) Thermal stabilities of sulphates: Thermal stabilities of sulphates also increase as the basic character of the metal hydroxide increases.

Anamalous behavious of Be: Be differs from the rest of the members of group

II   because of its

  (i)   Higher electronegativity

  (ii)   Smaller atomic and ionic radii

- Be shows diagonal relationship with Al since they have approximately the same polarizing power.
- Be and Mg crystalline in hcp, Ca & Sr in C_4 and Ba in bcc structures.
- The most abundant alkaline earth metal in the earth crust is Ca & least abundant is Ra.
- $BeCl_2$ has polymeric structure in the solid state but exists as a dimer in the vapour state.

Solid state

- $CaCl_2.6H_2O$ is widely used for melting ice on roads, particularly in very cold countries.
- Magnesium perchlorate ($Mg(ClO_4)_2$) is used as a drying agent under the name anhydrone.

BORON FAMILY

The elements

Boron	-	B (z=5)
Aluminium	-	Al (13)
Galium	-	Ga (31)
Indium	-	In (49)
Thalium	-	Tl (81)

Al is the third most abundant element found in the earth's crust after oxygen and silicon.

Physical Properties:

1. Electronic configuration $ns^2 np^1$: These elements are belonging to p-block of the periodic table.

2. Ionization energy: IE_1 values of IIIA are lower than the corresponding IIA group elements. On moving down the group from B to Al, IE_1 decreases as expected but the next Ga has slightly higher IE_1 than Al. It again decreases in In and increases in the last element Tl.

3. Oxidation states (O.S.): B and Al show an oxidation state +3 only while Ga, In an Tl show +1 and +3.

4. Inert pair effect: It is the reluctance of the s-electrons of the valance shell to take part in bonding. Inert pair effect increases from Be to Tl.

5. Atomic and ionic radii: Atomic and ionic radii is lower than II_A group elements due to greater nuclear charge of III_A group.

6. MP & BP: MP and BP is much higher than IIA group elements.

7. Electropositive Character (EP): The EP character first increases from B to Al and then decreases from Al to Tl. Thus, B is a typical non-metal and poor conductor of electricity. Al is a metal and is a good conductor of electricity.

8. Reducing Character: The reducing character decreases from Al to Tl as the electrode potential values for M^{+3} / M increases.

Chemical Properties:

1) Action of air: Crystalline B is unreactive only amorphous B shows reactivity.

$$4B + 3O_2 \rightarrow 2B_2O_3$$

Al reacts readily in air, even at ordinary temperature forming a protective layer of the oxide.

$$4\,Al + 3O_2 \rightarrow 2A_2O_3$$

Ga does not react with air. Tl is more reactive than Ga & In.

2) Action of water: Pure B doesn't react with water. Al decomposes boiling water evolving H_2.

$$2\,Al + 6H_2O \rightarrow 2Al\,(OH)_3 + 3H_2$$

Ga & In do not react with hot water. Tl is little more reactive than Ga.

3) Reactio with alkalies: B when fused with alkalies, dissolves to give borates, Al & Ga dissolve in concentration alkalies on heating forming metal aluminate and galate.

$$2\,B + 6\,NaOH \xrightarrow{fuse} 2\,Na_3BO_3 + 3H_2$$

$$Al + 2\,NaOH \xrightarrow{\Delta} 2\,NaAlO_2 + 3H_2$$

4) Reaction with N_2: When heated with nitrogen form nitrides.

$$2\,B + N_2 \xrightarrow{\Delta} 2BN \quad (Boron\ nitride)$$

5) Reaction with carbon: B & Al when heated with carbon form carbides.

$$4\,B + C \xrightarrow{\Delta} B_4\,C$$

$$4\,Al + 3C \xrightarrow{\Delta} Al_4C_3$$

6) Hydrides: These elements react with H_2 to form hydrides:

$$BF_3 + 3\,Li\,BH_4 \xrightarrow{Dry\,-o-} 2\,B_2H_6 + 3\,LiF$$

$$(Diborane)$$

B hydrides are quite stable but stability of the hydrates of the elements decreases from Al to Tl.

$$Al > Ga > In > Tl$$

Boron form number of hydrides such as B_2H_6 – diborane (6)

$$B_4H_{10} - \text{tetraborane (10)}$$
$$B_5H_9 - \text{pentaborane (9)}$$

The general formula of hydrides B_nH_{n+4} and B_nH_{n+6}.

B_2H_6 (Diborane) structure:

B atom undergoes Sp^3 hybridization.

Four B-H are normal covalent bonds, while two B-H-B are three centre electron pair bonds. These bonds are also called banana bonds.

Al forms a polymeric hydride of formula $(AlH_3)_x$. GaH_3 is even less stable while InH_3 and TlH_3 are extremely unstable.

7) Oxides: All the elements combine with oxygen when heated to form oxides. The formula of oxides M_2O_3.

Tl oxide is more stable, Tl_2O is more stable than Tl_2O_3 due to inert pair effect.

$$2\,Al(OH)_3 \xrightarrow{\Delta} Al_2O_3 + 3H_2O$$

Al_2O_3 (Alumina) exist in many crystalline forms depending upon its mode of preparation. The hardest alumina is called corundum. The other form γ-Al_2O_3 is called activated alumina.

Acid-base behaviours of hydroxides and oxides: On moving to down the group, acidic character decrease and increase the basic characters. B_2O_3 and $B(OH)_3$ are weakly acidic. Al_2O_3 & $Al(OH)_3$ are amphoteric, Ga_2O_3 and $Ga(OH)_3$ are also amphotesic. In_2O_3 and $In(OH)_3$ are slightly bases. Tl_2O_3 & $Tl(OH)_3$ are more basic. TlOH is strong base like alkali metal hydroxides.

8) Halides: All the elements react with halides to form trihalides of general formula MX_3 (where X = F, Cl, Br & I).

Boron trihalides	-	BF_3, BCl_3, BBr_3, BI_3
Al trihalides	-	AlF_3 ionic compound
		$AlCl_3$, $AlBr_3$, AlI_3 – covalent compounds

- B trihalides exists in monomers, AlX_3 exist as diner.
- Boron shows diagonal relationship with Silicon.
- B exhibits anomalous behavior because of its
  (i) small size (ii) higher electronegativity (EN), (iii) high IE

- B forms only covalent compound where as Al can form both covalent and ionic compounds.

    Eg: ahydrous $AlCl_3$ – covalent

           Hydrated $AlCl_3$ - Ionic

- Carboranes are derived from boranes by the replacement of BH^- units by CH units.
- Borazine (or) Borazole (or) Triborine triamine $(B_3N_3H_6)$ is colourless liquid and is also called inorganic benzene.
- Tl is highly toxic element.
- B shows maximum covalency 4 while Al shows six.
- Borax $(Na_2B_4O_7.10H_2O)$ is widely used in qualitative analysis.
- Androus $AlCl_3$ is used as a catalyst in Friedel – Crafts reaction and in cracking of petroleum.
- Due to their toxic nature, traces of Tl salts can cause loss of hair.
- Boric lotion is solution of boric acid used as an antiseptic and eye wash.
- Boron carbide B_4C is the hardest known artificial substances and is called Norbide.

CARBON FAMILY

The elements

Carbon	-	C (Z=6)
Silicon	-	Si (14)
Germanium	-	Ge (32)
Tin	-	Sn (50)
Lead	-	Pb (82)

These are collectively called as carbon family. Si is the second most abundant element after oxygen.

Physical properties:

1. Electronic configuration $ns^2npx^1npy^1$ (or) ns^2np^2

Elements of IVA are in fact p-block elements. Since the last electron in them is present in p-orbital.

2. Atomic radii:

The atomic radii increases from C to Pb due to the addition of new energy shell at each succeeding element.

$$C < Si < Ge < Sn < Pb$$

3. Ionization energy:

The IE1 of IVA elements are higher than those of the corresponding elements of group IIIA due to increased nuclear charge. IE decreases from C to Pb.

$$C > Si > Ge > Sn > Pb$$

4. Electronegativity:

Electronegativity decreases from C to Pb. C with an EN of 2.5 is the most EN element of IVA group.

$$C > Si > Ge > Sn > Pb$$

E.N   2.5   1.8   1.8   1.8   1.9

5. Oxidation states:

All the elements shows +4 oxidation state. As move from C to Pb +4 oxidation state decreases.

6. Allotropy:

If an element exists in 2 (or) more than two forms which have different physical properties but identical chemical properties, the phenomenon is called allotropy and different forms are called allotropes. Carbon exists in several forms which are originally regarded as amphous. These are coal, coke, charcoal etc. It also exist is two crystalline allotropic forms i.e. diamond and graphite. Pb do not shows allotropy.

(i) Diamond:

In diamond C is sp^3 hybridized. Each carbon thus forms covalent bonds with four other carbon atoms which lie at the corners of the regular tetrahedron. As a result diamond is a 3 dimensional network solid.

The C-C bond length is 1.54 A°. Diamond is the purest form of carbon. It is a bad conductor of electricity. Diamond is used for cutting glass, making borers for rock drilling and for making abrasives, precious gems & jewellery.

(ii) Graphite:

In graphite carbon in sp^2 hybridised. Each carbon is thus linked to three other carbon atoms forming hexagonal rings. Graphite has two dimensional sheet like structure. The distance between two layers 3.35A°. The C-C bond length is 1.42A°. Graphite is more stable than diamond.

Graphite is a soft black substance with metallic luster. It is a good conductor of electricity. Graphite is used as a lubricant and for making electrodes for dry cells. Mixed with desired quantities of clay, graphite is used for making lead pencils.

Si is known to exist in crystalline and amorphous forms. Ge exists in two crystalline forms while tin exist in different forms.

7. Catenation:

The property of self linking of atoms of an element through covalent bonds to form straight (or) branched chains and rings of different sizes is called catenation. The tendency for catenation decreases in the order:

$$C >>> Si > Ge \approx Sn >> Pb$$

8. M.P & B.P:

M.P & B.P are much higher than those of the IIIA group elements. M.P & B.P decreases from C to Pb due to size of the atoms increased and interatomic forces of attraction decreases.

$$C > Si > Ge > Sn > Pb$$

9. Density:

Density increases from C to Pb.

$$C < Si < Ge < Sn < Pb$$

Chemical Properties:

1. Nature of bonding:

The compounds of IVA group elements which show O.S. of +4 are covalent while those which show an oxidation state (O.S.) +2 are ionic nature. As move down the group, the tendency of the elements to form covalent compounds decreases but the tendency to form ionic compounds increases.

2. Formation of Complex:

The tendency to form complexes depends upon (i) small size of atom (ii) high charge (iii) availability of vacant d-orbitals of appropriate energy.

Carbon does not form complexes because of the absence of vacant d-orbitals.

$$SiF_4 + 2F^- \rightarrow [SiF_6]^{-2}$$

$$GeF_4 + 2F^- \rightarrow [GeF_6]^{-2}$$

$$PbCl_4 + 2Cl^- \rightarrow [PbCl_6]^{-2}$$

3. Formation of hydrates:

All the elements form covalent hydrides. The hydrides stability decrease from C to Pb. Carbon forms large number of hydrides called hydrocarbons. Si form less number of hydrides. These are called Silanes ($SinH_{2n+2}$) Ge forms hydride called Germanes. The reducing character of hydrides follows the order:

$$CH_4 < SiH_4 < GeH_4 < SnH_4 < PbH_4$$

4. Formation of halides:

(i) Dihalides:

All the elements form dihalides of formula MX_2 (where X=F, Cl, Br, I). The stability increases from C to Pb.

(ii) Tetrahalides:

All the elements form tetrahalides of formula MX_4. The stability decreases from C to Pb.

$$CCl_4 > SiCl_4 > GeCl_4 > SnCl_4 > PbCl_4$$

The order of thermal stability and volatility of tetrahalides.

$$MF_4 > MCl_4 > MBr_4 > MI_4$$

Tetrahalides of Ge, Sn & Pb behave as oxidizing agent and the oxidizing character of M^{+4} species increases in the order.

$$GeCl_2 < SnCl_2 < PbCl_2$$

Dihalides of Ge, Sn & Pb behave as reducing agent, the reducing character decreases in order:

$$GeCl_2 > SnCl_2 > PbCl_2$$

5. Formation of oxides:

All the elements of this group form two types of oxides (i) monoxides (ii) dioxides:

(i) Monoxide:

All the elements are form monoxide of the general formula MO (CO, SiO, GeO, SnO and PbO). The oxides except Sio and GeO are quite stable. Co is neutral while SnO and GeO are amphoteric.

Carbon monoxide:

It is neutral oxide, it is prepared in the laboratory by heating oxalic acid with concentration H_2SO_4. It's a poisonous gas, burns with blue flame and when inhaled produces suffocation and finally death. The CO combines with haemoglobin, carboxy haemoglobin is formed which destroys its capacily to supply oxygen to the body. It act as a reducing agent and is thus used to reduce heated metal oxides to metals.

$$Fe_2O_3 + 3Co \rightarrow 2Fe + 3CO_2$$

CO is a resonance hybrid of the following structures:

$$:C{=}\ddot{O}: \quad :C{\overset{+}{=}}\ddot{O}{:} \quad :C{\equiv}O^+$$

Because of the presence of lone pair of electrons. CO acts as hewis base and combines with many metals forming their respective metal carbonyls.

$$Fe + 5\,Co \rightarrow Fe\,(Co)_5$$

PbO: Red form of PbO is called litharge and yellow form is called massicot.

Dioxides:

All the elements form dioxides, general formula MO_2 (CO_2, SiO_2, GeO_2, SnO_2 & PbO_2).

CO_2:

It is monomeric, linear molecule and hence exists as a gas. It is prepared by the action of dilution HCl on calcium. It is used in fire extinguishers and in the manufacture of aerated water.

Dry ice:

Solid carbondioxide is called dry ice. It is used as coolant for preserving perishable articles in food industry and making cold baths in the laboratory.

Silica (SiO_2):

Silicon dioxide (or) silica is solid at room temperature and has three dimensional structures. It's hybridization is SP^3. The structure of SiO_2 is highly stable and thus silica is very hard and has high melting point.

Quartz is crystalline form of silica.

Pb forms Pb_3O_4 (red lead or sindhur). Plumbo solvency is the dissolution of Pb in water containing air & CO_2 forming soluble $Pb(OH)_2$ which gives highly poisonous Pb^{+2} ions.

6) Formation of carbides:

Compounds of carbon with elements of lower or equal EN are called carbides. Carbides are 3 types:

(1) Ionic carbides: These are compounds of carbon with more EN.

Eg: Be_2C, CaC_2, Al_4C_3 etc.

(2) Covalent carbides: These are compounds of carbon with slightly EN less than that of carbon.

$$Eg: B_4C, Sic \ etc.$$

(3) Interstitial carbides: These are compounds of carbon with transition element.

$$Eg: TiC, WC, ZrC \ etc.$$

Carbides are prepared by heating the element.

$$Si + C \xrightarrow{\ 2500\,K\ } SiC$$

$$SiO_2 + 3C \xrightarrow{\ 2300\,K\ } SiC + 2CO; \quad CaO + 3C \xrightarrow{\ 2300\,K\ } CaC_2 + 2CO$$

CaC_2 is used for preparation of acetylene. Silicon carbide is very hard and is used as abrasive under the name carborandum. Tungsten Carbide (WC) is used in making tools & dies.

7) Silicates:

Rocks, clays and soils are made up of silicates of Al, Fe, Mg and other metals. Albestors $(CaSiO_3.3MgSiO_3.H_2O)$ is a double chain silicates. Mica, clay and talc are the examples of the sheet silicates.

8) Glass:

Glass is a transparent of translucent super cooled complex mixture of non-crystalline silicates. Soft glass (or) Soda glass is mixture of Na & Ca silicates with composition as $Na_2O.CaO.6SiO_2$. Pyrax glass (or) corning glass is a mixture of Na & Al borosilicates. Flint glass is a mixture of K & Pb silicate Na & Al borosilicates. Flint glass is a mixture of K & Pb silicate.

Common glass is soluble in HF due to the formation of soluble product.

$$Na_2SiO_3 + 8\,HF \rightarrow 2\,NaF + H_2SiF_6 + 3H_2O$$

$$CaSiO_3 + 8\,HF \rightarrow CaF_2 + H_2SiF_6 + 3H_2O$$

9) Silicones:

Silicones are synthetic organosilicon compound. These are used for making water proof papers, wool, textiles, wood etc. as lubricants.

10) Gaseous fuels of common use:

(1) Water gas ($CO+H_2$) is a mixture of CO and H_2 and is obtained by passing steam over red hot coke. It is used as an industrial fuel.

$$C + H_2O \xrightarrow{473-1273\ K} \underline{CO + H_2}$$

(2) Producer gas ($CO+N_2$): It is a mixture of CO and N_2 and is obtained by passing air over heated coke. It issued as an industrial fuel.

$$2C + O_2 + 4N_2 \longrightarrow \underline{2CO + 4N_2}$$

Producer gas

Carbogen is a mixture of 95% O_2 & 5% CO_2. It is used for artificial respiration for victims of co-poisoning.

(3) Carburetted water gas:

It is a mixture of H_2 (34-38%), CO (23-28%), saturated hydrocarbons (17-21%), unsaturated hydrocarbons (13-16%), CO_2 (2%) and N_2 (5%).

(4) Semi water gas:

It is mixture of CO, H_2 and N_2 with CO_2 & CH_4.

(5) Oil gas – CH_4 + H_2 + CO + CO_2.

(6) LPG – It is a mixture of butane and propane.

(7) Biogas or Gobar gas – mixture of CH_4 + H_2 + CO_2 + N_2.

(8) Coal gas – It is mixture of CH_4 + H_2 + CO + C_2H_2 + C_2H_4 + N_2 + CO_2.

(9) Natural gas – It is mainly CH_4 with C_2H_6, H_2 and rest CO + CO_2.

- Graphite is oxidized by concentration HNO_3 to give graphitic acid ($C_{11}H_4O_5$).

- Diamond on heating at 1800-2000°C gives graphite.
- Tetraethyl lead (TEL), $Pb(C_2H_5)_4$, is prepared by the action of ethyl chloride on sodium lead alloy.
- It is used as high anti knock agent to improve the quality of gasoline.
- SiC is used as high temperature semiconductor in transistor diode rectifiers.
- Quartz is the most common and purest variety of SiO_2.

NITROGEN FAMILY

The elements

Nitrogen	-	N (z=7)
Phosphorus	-	P (15)
Arsenic	-	As (33)
Antimony	-	Sb (51)
Bismuth	-	Bi (83)

These constituent VA group of the periodic table.

Physical Properties:

1. Electronic configuration – nS^2nP^3

The elements of this family belong to p-block.

2. Atomic radii:

Atomic radii of these elements increase from N to Bi.

$$N < P < As < Sb < Bi$$

3. Ionization energy:

IE of these elements is much higher than the corresponding elements of group IVA. IE decrease from N to Bi due to increase atomic radii.

$$N > P > As > Sb > Bi$$

4. Electronegativity:

EN of these elements decreases from N to Bi due to increase in the atomic radii.

$$N > P > As > Sb > Bi$$

EN: 3.0 2.1 2.0 1.9 1.9

5. Melting and Boiling point:

M.P & B.P of these elements first increase from N to As due to increase in their atomic size & then decrease to Sb and Bi because of their tendency to forms three covalent bonds instead of five due to inert pair effect.

$$N < P < As < Sb < Br$$

6. Oxidation State:

(1) Negative oxidation state: The elements of this group contain five electron in their valency shell and hence can acquire the nearest gas configuration by gaining 3 more electron to form triply changed –ve ions. Such as N^{3-} and phosphides P^{-3} ion. The tendency of these elements to show -3 O.S. decrease from N to Bi.

(2) Positive Oxidative on State: The elements of VA group show +3 & +5. The stability of +3 oxidation state increases. Nitrogen shows -1, -2, -3, +1, +2, +3, +4 and +5 oxidation states in NH_2OH, N_2H_4, NH_3, N_2, N_2O, NO, N_2O_3, N_2O_4 & N_2O_5 respectively.

7) Non-metallic & metallic character:

As we move down from N to Bi metallic character increases.

$$\left. \begin{array}{c} N \\ \\ P \end{array} \right\} \text{Non-metals}$$

$$\left. \begin{array}{c} As \\ \\ Sb \end{array} \right\} \text{Metalloids}$$

Bi - Typical metal

8) Electrical & Thermal conductivities

of group VA increases from N to Bi due to decrease in IE N & P are non-conductors but Bi is an excellent conductor.

9) Catenation:

N_2 has little tendency for catenation since N-N single bond is weak due to repulsion between non-bonded electron pairs in a smaller nitrogen atom. The bond energy of P-P bond is very large. Phosphorus has a distinct tendency for catenation forming cyclic as well as open chain. The element bond energies decrease rapidly.

$$P > N > As > Sb > Bi$$

10) Elemental State:

Because of small size and high EN, N_2 has a strong tendency to form multiple ($P\pi$-$P\pi$) bonds. Thus, N_2 exists as a diatomic gas in which two N atoms are linked by a triple bond. P, As & Sb exist as tetra atomic tetrahedral molecules i.e. P_4, As_4 & Sb_4. Bi has metallic bonding.

11) Allotropy:

All the elements except Bi show allotropy. Two solid forms of nitrogen i.e. α-nitrogen & β-nitrogen. Phosphorus exists in a number of allotropic forms such as white (or) yellow, red, scarlet, black & violet phosphorus. As exists in 2 allotropic forms they are yellow & grey arsenic.

White P when exposed to light aquires yellow colour. Red P consists of a complex chain structure. Black P has layer structure. It is most stable form and is a good conductor of electricity. It has a metallic luster.

12) Nature of bonding:

All the elements form covalent compounds by sharing of electrons.

Chemical Properties:

1) Formation of hydrides:

All the elements of group VA form volatile hydrides of the type MH_3.

Ammonia (NH_3), phosphine (PH_3), arsine (AsH_3), stilbine (SbH_3) and bismuthine (BiH_3).

(i)   Thermal stability of these hydrates decreases gradually form NH_3 to BH_3.
$$NH_3 > PH_3 > AsH_3 > SbH_3 > BiH_3$$

The reason is the strength of the M-H bond decreases due to an increase in the atomic size of the element.

(ii) Reducing character of these hydrates increase from NH_3 to BiH_3.

$$NH_3 < PH_3 < AsH_3 < SbH_3 < BiH_3$$

(iii) Basic Character: Basic character decreases from NH_3 to BiH_3.

(iv) Hydrogen bonding: Due to small size & high EN of nitrogen, NH_3 form H-bonds but other hydrides of this group donot exhibit H-bonding.

(v) M.P & B.P: The M.P & B.P of the hydrides of the element increases.

The increasing order of M.P   - $PH_3 < AsH_3 < SbH_3 < NH_3$

The increasing order of B.P   - $PH_3 < AsH_3 < NH_3 < SbH_3$

(vi) Bond angles: The hydrides of these elements have pyramidal shape. Bond angles gradually decrease due to decrease in bond pair-bond pair repulsion.

$$NH_3 > PH_3 > AsH_3 > SbH_3 > BiH_3$$

(vii) Dipole moments: Dipole moments decreases from NH_3 to BiH_3.

$$NH_3 > PH_3 > AsH_3 > SbH_3 > BiH_3$$

The hydrates can be prepared by the following reactions.

$$Ca_3P_2 + 6H_2O \longrightarrow 2PH_3 + 3\ Ca(OH)_2$$

$$P_4 + 3KOH + 3H_2O \longrightarrow PH_3 + 3KH_2PO_2$$

2) Formation of halides:

a) Trihalides: All the elements form halides of the type MX_3.

$$NCl_3 + 3H_2O \longrightarrow 3HClO + NH_3$$

$$PCl_3 + 3H_2O \longrightarrow H_3PO_3 + 3HCl$$

The hydrolysis decreases from NCl_3 to $BiCl_3$

$$NCl_3 > PCl_3 > AsCl_3 > SbCl_3 > BiCl_3$$

Basic nature:

Due to presence of line pair of electrons, the trihalides act as Lewis-bases. Increasing order of Lewis base strength of trihalides is $NF_3 < NCl_3 < NBr_3 < NI_3$.

Lewis acid strength   - $PF_3 > PCl_3 > PBr_3 > PI_3$

b) Pentahalides:

P, As & Sb also form pentahalides. 'N' does not form pentahalides due to the absence of d-orbitals in the valence shell. The pentahalides have trigonal bipyramid shapes.

3) Oxides:

All the elements of this group form two types oxides i.e M_2O_3 & M_2O_5 and are called trioxides and pentoxides. The trioxides of N, P & As are acidic. The acidic order decreases in the order.

$$N_2O_3 > P_2O_3 > As_2O_3$$

The Sb and Bi are basic in character. All the pentoxides are acidic but the acidity decreases from N to Bi.

$$N_2O_5 > P_2O_5 > As_2O_5 > Sb_2O_5 > Bi_2O_5$$

The acidic strength of oxides of nitrogen in the order.

$$N_2O < NO < N_2O_3 < N_2O_4 < N_2O_5$$

N_2O_3 -   acidic nature, it is anhydride of nitrous acid.

NO_2 -   coloured gas paramagnetic nature

N_2O_4 -   Diamagnetic nature, it's oxidizing agent

N_2O_5 -   a colourless solid, it's an anhydride of nitric acid acts as excellent oxidizing agent.

4) Formation of oxy acids:

(a) Oxy acids of nitrogen:

Oxy acid	Formula	Structure	Acidity
Hyponitrous acid	$H_2N_2O_2$	HO-N=N-OH	Weak dibasic acid
Nitrous acid	HNO_2	HO-N=O	Weak acid, unstable
Nitric acid	HNO_3	HO-N=O	Strong dibasic acid
Pernitric acid	HNO_4	HO-O-N=O	Unstable acid

(B) Oxy acids of phosphorus:

	Oxy acid	Formula	Structure	O.S.	
1	Hypophosphorus acid	H_3PO_2		+1	Monobasic acid
2	Phosphorous acid	H_3PO_3		+3	Dibasic acid
3	Orthophosphoric acid	H_3PO_4		+5	Weaktribasic acid
4	Hypophosphoric acid	$H_4P_2O_6$		+4	Tetrabasic acid
5	Pyrophosphoric acid	$H_4P_2O_7$		+5	Tetrabasic acid
6	Metaphosphoric acid	HPO_3		+5	Monobasic acid
7	Phosphorous acid	H_3PO_3			Dibasci acid
8	Peroxy phosphoric acid	H_3PO_5			Tribasic acid

The decreasing order of acid strength

$$HNO_3 > H_3PO_4 > H_3AsO_4 > H_3SbO_4$$

Anamalous behavior of nitrogen:

N_2 difers from rest of the members of its family because of (i) small size (ii) high EN & (iii) absence of d-orbitals in the valence shell.

Phosphorescence:

White or yellow phosphorus glows in dark due to its slow combustion in air. The energy of combustion is emitted as light.

- The disease caused by the constant touch with white phosphorus is called phossy jaw.
- Tetraphosphorus trisulphide, P_4S_3 is used in strike any where matches.
- Hydrazine (N_2H_4) acts as oxidizing agent like H_2O_2. Hydrazine and its derivatives are used as rocket fuels.
- Phosphorous Pentoxide (P_4O_{10}) due to its appearance as a snowy powder is called flowers of phosphorus.
- Arsenic trioxide, As_4O_6 is called white arsenic and is poison.
- Amatol which is used as an explosive is a mixture of NH_4NO_3 (80%) & TNT (20%).
- In tooth paste, $CaHPO_4.2H_2O$ is added as a mild abrasive and polishing agent.
- Proteins the building blocks of our body contain 16% of nitrogen in them.
- BiOCl is called pearl white.
- Thomas slag – mixture of $Ca_3(PO_4)_2$ & $CaSiO_3$ and it is used as fertilizer.
- Tartaremetic is potassium antimonyl titrate which contains antimony and is used in medicine as antipoison.
- Nitric acid stains skin yellow due to the formation of a nitro compound called xanthoprotein.

OXYGEN FAMILY

The elements

Oxygen	-	O (z=8)
Sulphur	-	S (16)
Selenium	-	Se (34)
Tellurium	-	Te (52)
Polonium	-	Po (84)

Constitute VIA group of the periodic table. The first four members of this group are also collectively called chalcogens.

Physical Properties:

1. Electronic configuration – $ns^2 np^4$ (or) $ns^2 np_x^2 np_y^1 np_3^1$

2. Atomic & ionic radii:

Atomic and ionic radius increases from O to Po due to increase in the number of electron shells.

$$O < S < Se < Te < Po$$

3. Ionization energy:

The IE_1 of the elements of VIA are lower than those of the corresponding elements of VA group. The IE decreases from O to S and then fall regularly from S to Te due to increase in their atomic radii & shielding effect.

$$O < S < Se < Te < Po$$

4. Electronegativity:

Elements of VIA group have higher values of EN than the corresponding elements of VA group, EN decrease gradually from O to Po.

$$O > S > Se > Te > Po$$

EN   3.5 2.5 2.4   2.1 2.0

5) Electron affinities (EA):

EA decrease from S to Po, Oxygen has low EA.

$$O > S > Se > Te > Po$$

6) Oxidation States:

1) Negative Oxidation state (O.S.)

Oxygen because of its high EN & IE shows an O.S. of -2. The -2 oxidation state decreases from O to Po. In oxygen peroxides an O.S. of -1 is shown.

2) Positive Oxidation State (O.S.)

These elements show positive oxidation states +2, +4 and +6. Oxygen shows only +2 (O.S.) (OF_2) but does not show +4 & +6 due to absence of d-orbitals in its valence shell. Oxygen behaves as a divalent element only. Remaining all elements shows +4 and +6 O.S. because of presence of d-orbitals in valence shell. In group the stability of +4 O.S. increases from O to Po due to inert pair effect.

7) Non-metallic and metallic character:

There is a gradual change from non-metallic to metallic character as we move down the group due to decrease in IE down the group from O to Po.

O and S	-	non-metals
S and Te	-	metalloids
Po	-	metallic

8) Nature of bonding in compounds:

The compounds of oxygen with more EN are ionic. S, Se & Te because of low EN show more covalent character in majority of their compounds.

9) Molecular structure – atomicity:

O has a strong tendency to form multiple bonds and hence exists as a diatomic O_2 ($O=O$) gas. While S, Se, Te because of bigger size have little or no tendency to form $P\pi$-$P\pi$ multiple bonds, but prefer to form single bond. Thus these elements exist as Octa-atomic i.e. S_8, Se_8, Te_8. The decreasing tendency to exist in puckered 8-membered ring structure is

$$S > Se > Te > Po$$

In S_8 molecule, each sulphur atom undergoes Sp^3-hybridisation.

10) Catenation:

Oxygen has some but S has greater tendency for catenation. The decreasing order of catenation from O to Te.

$$S > Se > O > Te$$

11) Allotropy:

All the elements show allotropy. O exists in two non-metallic forms i.e O_2 & O_3. Sulphur exists in several forms i.e rhombic, monoclinic and plastic sulphur, rhombic and monocyclic sulphur have S_8 ring structure while plastic sulphur consists of inverted long chains.

Se	-	red (non-metallic), grey (metallic)
Te	-	exists in two forms
Po	-	α & β (both metallic)

12) Reactivity of elements:

The general reactivity decreases from O to Po. Oxygen is slightly less reactive than halogens and reacts with all elements usually at high temperature.

Chemical Properties:

1) Formation of hydrides:

All these elements form volatile, stable, bivalent hydrides of the formula H_2M i.e. H_2O, H_2S, H_2Se, H_2Te & H_2Po. The central atom in these hydrides is Sp^3 hybridized.

(i) M.P & B.P:

The M.P & B.P of hydride of oxygen are abnormally high due to strong H-bonding which is attributed to high EN & small size of oxygen. The M.P & B.P of hydrides gradually increase from H_2O to H_2Te.

$$H_2O > H_2Te > H_2Se > H_2S$$

H_2O has the highest and H_2S has the lowest values of M.P. & B.P.

(ii) Volatility

The volatility increase from H_2O to H_2S and then decreases from H_2S to H_2Te.

$$H_2O < H_2Se < H_2Te < H_2S$$

(iii) Thermal stability of the hydrides decrease from H_2O to H_2Te.

$$H_2O > H_2S > H_2Se > H_2Te$$

(iv) Reducing character:

Hydrides of all these elements except that of oxygen i.e H_2O behave as reducing agents. The reducing character increase from H_2S to H_2Te.

$$H_2S < H_2Se < H_2Te$$

(v) Acidic Character:

The hydrides of these elements behave as weak acids their acidic strength increase from H_2O to H_2Te.

(vi) Covalent Character:

It goes on increasing with atomic number. This can be explained on the basis of Fajan's rules.

$$H_2O < H_2S < H_2Se < H_2Te$$

(vii) Bond angles:

The bond angles decrease from H_2O to H_2Te due to decrease is the EN of the central atom.

$$H_2O > H_2S > H_2Se > H_2Te$$

$$(104°) \ (92°) \ (91°) \ (90°)$$

(viii) Dipole moments:

The dipole moments of H_2M decrease as the EN of the central atom decreases down the group. The dipole moment decrease from H_2O to H_2Te.

$$H_2O > H_2S > H_2Se > H_2Te$$

2) Formation of halides:

(a) Halides of Oxygen:

Since O doesn't contain d-orbitals in its valence shell, it cannot expand its octet. As a result, oxygen halides are limited to maximum no. of two halogens linked one oxygen i.e. oxygendifluoride (F_2O), chlorine monoxide (Cl_2O) and bromine monoxide (Br_2O). All these halides have Sp^3 hybridisation.

$$X_2O + H_2O \rightarrow 2 \ HOX \ (X = Cl \ or \ Br)$$

(b) Halides of S, Se, Te and Po:

These elements form a variety of halides.

(i) Hexahalides:

All the elements except oxygen form hexafluorides i.e SF_6, SeF_6 and TeF_6. The stability of these hexafluorides decreases from SF_6 to TeF_6. SF_6 is practically inert, slightly more reactive. TeF_6 is hydrolysed by water.

$$SF_6 < SeF_6 < TeF_6$$

(ii) Tetrahalides:

All the elements except oxygen form tetrafluorides (SF_4, SeF_4, TeF_4) and tetrachlorides (SCl_4, $SeCl_4$, $TeCl_4$). These have trigonal bipyramid (Sp^3d) geometry.

The stability of tetrachlorides increases from SCl_4 to $TeCl_4$.

(iii) Dihalides:

Only S forms a dichloride i.e. SCl_2. It has tetrahedral geometry but bent structure due to presence of lone pair of electron on S atom.

(iv) Monohalides:

Only monochlorides and monobromides of S and Se are known. They exist as dimmers i.e. S_2Cl_2, Se_2Cl_2 etc.

3) Fomation of oxides:

These elements form a variety of oxides in different oxidation state from +2 to +6.

(i) Acidic character of oxides of these elements decreases from S to Po.

$$SO_2 > SeO_2 > TeO_2 > PoO_2$$

$$SO_3 > SeO_3 > TeO_3$$

(ii) Acidic character of oxides of a particular element increases with the increase in oxidation number of the central element.

(iii) The oxides of S are more stable than the corresponding oxides of selenium.

(a) Sulphur dioxide:

SO_2 is a gas and forms discrete molecules even in the solid state. It is acidic nature and is also called anhydride of sulphurous acid (H_2SO_3). It can act as reducing agent & oxidizing agent. SO_2 acts as Lewis base due to the presence of a lone pair of electrons on S atom.

SO_2 molecule has a bent structure with O-S-O bond angle of 119°.

Gaseous SeO_2 has the same structure but the solid form consists of infinite non-planar chains.

(b) Sulphur trioxide (SO_3):

It is an acidic oxide and is regarded as an anhydride of H_2SO_4. It acts as an oxidizing agent. In the gas phase it exists as planar triangular molecular species involving Sp^2 – hybridization.

The structure of solid sulphur trioxide is complex. It possesses either cyclic trimer structure or an infinite helical chain made up of linked SO_4 tetrahedron.

4) Oxy acids of Sulphur:

Name	Formula	Structure	S oxidation state
Sulphurous acid	H_2SO_3		+4
Thiosulphurous acid	$H_2S_2O_2$		+4
Sulphuric acid	H_2SO_4		+6
Thiosulphuric acid	$H_2S_2O_3$		+6, +2
Dithionous acid	$H_2S_2O_4$		+3
Pyrosulphuric acid (or) Oleum	$H_2S_2O_7$		+6
Dithionic acid	$H_2S_2O_6$		+5
Peroxymonosulphuric acid (or) Caro's acid	H_2SO_5		+6
Peroxy disulphuric acid (or) Marshall's acid	$H_2S_2O_8$		+6

Anamolous behavior of oxygen

Oxygen, the first element of VIA group differs from rest of its family members due to small size, high EN, high IE and non-availability of d-orbitals in the valence shell.

HALOGEN FAMILY

The elements

Fluorine	-	F (Z=9)
Chlorine	-	Cl (17)
Bromine	-	Br (35)
Iodine	-	I (53)
Astaline	-	At (85)

Constitute group VIIA of the periodic table.

Physical Properties:

1. Electronic configuration: nS^2nP^5

These elements are p-block elements and contain seven electrons in their respective valence shells.

2. Atomic and ionic radii:

A halogen atom has the smallest radius as compared to other elements in its period. The atomic radii increase from F to I due to increase in the number of shells.

Ionic radii also increase regularly from F to I.

$$F < Cl < Br < I$$

3. Ionization energy:

IE is higher than those of the corresponding elements of group VIA. IE decrease from F to I.

$$F > Cl > Br > I$$

Thus I which has a comparatively low value of IE has a tendency to lose an electron to form positive iodinium ion.

4. Electronegativity:

F is most EN element in the periodic table. With increase in atomic number down the group, the EN decreases.

$$F > Cl > Br > I$$

$$EN \quad (4.0)\,(3.2)\;(3.0)\;(2.7)$$

5. Electron affinity:

Electron affinity of chlorine, bromine and iodine decreases as the size of the atom increases. The decreasing order of electron affinity

$$Cl > F > Br > I$$

The Cl has the highest electron affinity.

6. Oxidation state:

All the halogens show an oxidation state of -1. While other halogens also show positive oxidation states up to a maximum of +7, due to the availability of vacant d-orbitals in the valence shell of these atoms.

7. Nature of bonds:

All the elements have seven electrons in the valence shell and hence require one more electron to acquire the nearest inert gas configuration. The halides of highly EP metals are ionic while those of weakly EP metals and non-metals are covalent. The tendency to form ionic compounds decreases from F to I.

8. Non-metallic character:

All the halogens are non-metallic in nature due to their higher IE. The non metallic character gradually decrease from F to I.

9. Atomicity and Physical State:

All the halogens exist as diametic covalent molecules. F_2 & Cl_2 are gases at room temperature, Br_2 is corrosive liquid and I_2 is volatile solid.

10. Colour:

All the halogens have characteristic colours:

F_2	-	light yellow
Cl_2	-	greenish yellow
Br_2	-	reddish brown
I_2	-	deep violet

11. Bond dissociation energy:

Bond dissociation energy increases from F_2 to Cl_2 and decreases from Cl_2 to I_2.

$$F_2 < Cl_2 > Br_2 > I_2 \text{ (or) } Cl_2 > Br_2 > F_2 > I_2$$

12. Bond length:

Bond length increase from F_2 to I_2 due to increase the size of halogen atom.

$$F_2 < Cl_2 < Br_2 < I_2$$

13. M.P & B.P:

M.P & B.P of these elements increase from F_2 to I_2 due to an increase in the Vander Waals forces.

$$F < Cl < Br < I$$

14. Solubility:

Halogens, being non polar in nature do not readily dissolve in polar solvent like water. F_2 reacts with H_2O vigorously even at low temperature forming a mixture of O_3 & O_2.

$$2F_2 + 2H_2O \rightarrow 4HF + O_2$$

$$3F_2 + 3H_2O \rightarrow 6HF + O_3$$

Cl_2, Br_2 & I_2 are more soluble in organic solvents like CCl_4, CS_2 or $CHCl_3$ and produced coloured solutions.

15. Oxidising power:

All the halogens act as strong oxidizing agents. The oxidizing power, decreases from F to I.

$$F_2 > Cl_2 > Br_2 > I_2$$

16. Heat of hydration:

The heat of hydration of the halide ion decreases as the size of the halogen decreases down the group from F to I.

$$F^- > Cl^- > Br^- > I^-$$

Chemical Properties:

1. Reactivity:

All the halogens are chemically very reactive elements. This is due to their low dissociation energy. The reactivity decrease in the order:

$$F > Cl > Br > I$$

2. Formation of halides:

Halogens combine with all the elements except He, Ne and Ar forming a large number of binary halides.

a) Halides of metals:

(i) There is a regular gradation from ionic to covalent bonding as the atomic no. of the halogen increases for the same metal atom.

The ionic character of M-X bond and M.P & B.P of halides decrease in the order M-F > M-Cl > M-Br > M-I.

(ii) Metals of low IE such as alkali metals form ionic halides where as metals with high IE such as transition metals form covalent halides. Molecular halides show decrease in M.P & B.P as MI > MBr > MCl > MF.

(iii) Halides of metals in their higher O.S. are more covalent than those formed in lower O.S.

AgX shows the solubility trend

$$AgI < AgBr < AgCl < AgF$$

b) Halides of non-metals:

(i)   Halides of non-metals are generally covalent in nature.

(ii)   The strength of M-X bond for a particular non-metal decreases in the order M-F > M-Cl > M-Br > M-I, bond strength of H-X decreases from HF to HI.

$$H-F > H-Cl > H-Br > H-I$$

3. Hydrides:

All the halogents combine directly with H_2 to form halogen acids but their reactivity progressively decreases from F to I.

$$H_2 + X_2 \quad \rightarrow \quad 2HX$$

(i) B.P (or) Volatility:

The B.P of halide hydrides gradually increase from HF to HI.

$$HCl < HBr < HI < HF$$

Volatility decreases in the order HCl > HBr > HI > HF.

(ii) Thermal stability:

Thermal stability of the hydrides decreases from HF to HI.

$$HF > HCl > HBr > HI$$

(iii) Acidic Strength:

The acidic strength of halogen acids decreases from HI to HF.

$$HI > HBr > HCl > HF$$

The conjugate base strength of these acids increases in the order:

$$I^- < Br^- < Cl^- < F^-$$

(iv) Reducing properties:

The reducing properties, increase in the order because stability of hydrides decreases.

$$HF < HCl < HBr < HI$$

(v) Dipole moment:

The dipole moments of HX decrease in the order halogen atom decreases from F to I.

4. Oxides:

Halogens do not combine readily with oxygen. A number of compounds of halogens with O_2 have been prepared by indirect method:

F_2	-	Oxygendifluoride (OF_2)
		Oxygen fluoride (O_2F_2)
Cl_2	-	Cl_2O, ClO_2, Cl_2O_6, Cl_2O_7
I_2	-	I_2O_5
Br	-	Br_2O, BrO_2, BrO_3

All the oxides of halogens are powerful oxidizing agents.

5. Oxoacids of halogens:

F_2 does not form any axoacid since it is the strongest oxidizing agent.

Oxo acids are of four types. They are hypohalous acid (HXO) halous acid (HXO_2), halic acid (HXO_3) and perhalic acid (HXO_4).

51

Oxidation state	Chlorine	Bromine	Iodine	Thermal stability & acid strength	Oxidising power
+1	HClO	HBrO	HIO		
+3	$HClO_2$	--	--	*Increases* ↓	*Decreases* ↓
+5	$HClO_3$	$HBrO_3$	HIO_3		
+7	$HClO_4$	$HBrO_4$	HIO_4		

Acidity decreases →

(i) Hybridized ion:

The halogen atom is Sp^3 hybridized.

(ii) Acidic character:

All these acids are monobasic containing an –OH group. The acidic character of the oxoacids increases with increase in O.S. $HClO < HClO_2 < HClO_3 < HClO_4$

The strength of the conjugate bases of these acids follows:

$$ClO^- > ClO_2^- > ClO_3^- > ClO_4^-$$

(iii) Oxidising power and thermal stability:

The oxidizing power of these acids decreases as the oxidation number increases.

$$HClO > HClO_2 > HClO_3 > HClO_4$$

due to O.S. increases, the halogen – oxygen bond becomes more covalent. As a result thermal stability increases.

$$HClO < HClO_2 < HClO_3 < HClO_4$$

The increasing stability order of anions of oxoacids of Cl_2 is

$$ClO^- < ClO_2^- < ClO_3^- < ClO_4^-$$

(iv) Perhalates are strong oxidizing agents, the oxidizing power is in the order:

$$BrO_4^- > IO_4^- > ClO_4^-$$

(v) The acidity of oxoacids of different halogens having the same oxidation number decrease with increase in the atomic size of the halogen.

$$HClO_4 > HBrO_4 > HIO_4$$

6. Interhalogen compounds:

The compounds of one halogen with the other are called interhalogens compounds. The main reason for their formation is the large EN & size difference between the different halogens.

AB	AB$_3$	AB$_5$	AB$_7$
ClF	ClF$_3$, BrF$_3$	BrF$_5$, IF$_5$	IF$_7$
BrF, BrCl, ICl, IBr, IF	IF$_3$, ICl$_3$		

General Properties:

(1) Largest halogen always serves the central atom.
(2) The highest inter halogen compound Eg: IF$_7$.
(3) The bonds in inter halogen compounds are essentially covalent.
(4) They ionize in solution (or) in the liquid state.

$$2 ICl \leftrightarrows I^+ + ICl_2^-$$

(5) They are strong oxidizing agents.
(6) Structure:
Inter halogen compounds
AB type   -   linear

AB_3 type   -   distorted trigonal bipyramidal (dSp^3-hybridization)

AB_5 type   -   distorted octahedral (d^2Sp^3)

AB_7 type   -   pentagonal bipyramidal

(7) Reaction with alkalies:

With cold and dilute NaOH, F_2 gives OF_2 while with hot and concentrated NaOH, it gives O_2.

$$2F_2 + 2\ NaOH \xrightarrow{Cold} 2NaF + OF_2 + H_2O$$

$$2F_2 + 4\ NaOH \xrightarrow{hot} 4\ NaF + 2H_2O + O_2$$

Other halogens

$$2\ NaOH\ (dil) + X_2 \xrightarrow{Cold} NaXO + NaX + H_2O$$

$$6\ NaOH\ (conc.) + 3\ X_2 \xrightarrow{hot} NaXO_3 + 5\ NaX + 3H_2O$$

(8) Polyhalide ions:

Halogens (or) interhalogens combine with halide ions to form polyhalide ions. Many other examples of polyhalides ions are:

Ions	Hybridisation	Structure	One pair of electron
Cl_3^-, Br_3^-, ICl_2^-, IBr_2^-, I_3^-	Sp3d	Linear	3
Cl_3^+, Br_3^+, I_3^+, ICl_2^+, IBr_2^+	Sp^3	Bent	2
ICl_4^-, BrF_4^-, I_5^-	Sp^3d^2	Squar planar	2
ICl_4^+, BrF_4^+, I_5^+	Sp^3d	Distorted tetrahedral	1
I_7^-, IF_6	Sp^3d^3	Distorted octahedral	1
I_7^+	Sp^3d^2	Octahedral	-

(9) Anamlous behavior of F_2:

F$_2$ differs from rest of the elements of its family due to

(a) Its small size (b) highest EN (c) low bond dissociation energy and (d) absence of d-orbitals in the valence shell.

(10) Chromyl chloride test:

When solid chloride is heated with concentration H_2SO_4 in presence of solid $K_2Cr_2O_7$ in a dry test tube, deep red vapours of chromyl chloride are evolved.

$$NaCl + H_2SO_4 \rightarrow NaHSO_4 + HCl$$

$$K_2Cr_2O_7 + 2H_2SO_4 \rightarrow 2KHSO_4 + 2CrO_3 + H_2O$$

$$CrO_3 + 2HCl \rightarrow CrO_2Cl_2 \uparrow + H_2O$$

(Chromyl chloride)

(ii) Preparation of pure Cl$_2$:

Pure Cl$_2$ may be obtained by heating dry plantinic chloride (PtCl$_4$) (or) gold chloride (AlCl$_3$) in hard glass tube.

$$PtCl_4 \xrightarrow{374°C} Pt\,Cl_2 + Cl_2 \xrightarrow{582°C} Pt + 2Cl_2$$

$$2\,All\,Cl_3 \xrightarrow{175°C} 2\,AuCl + 2Cl_2 \xrightarrow{185°C} 2Au + 3Cl_2$$

12. Iodine turns starch paper blue. The blue colour is due to the formation of starch – iodine complex. Similarly Br$_2$ turns starch paper yellow.

- HF reacts with glass to form sodium and calcium fluosilicate Na_2SiF_6 (or) $CaSiF_6$. So, it is used for etching of glass.

$$Na_2SiO_3 + 6HF \rightarrow Na_2SiF_6 + 3H_2O$$

$$CaSiO_3 + 6\,HF \rightarrow CaSiF_6 + 3H_2O$$

- Halogen react with NH$_3$ and give different product.

$$8\,NH_3 + 3\,Cl_2 \rightarrow N_2 + 6\,NH_4Cl$$

(excess)

$$NH_3 + 3\,Cl_2 \rightarrow NCl_3 + 3\,HCl$$

- Freons are chlorofluoro carbons (CFCs). Freon-II is CCl_3F. Freon-12 is CCl_2F_2. It is used as a refrigerant.
- Aqueous solution of NaOCl is called javelle water and is used as bleaching agent.
- Tincture of iodine is a mixture of I_2 & KI dissolved in rectified spirit.
- Iodex ointment contains iodoform which liberates I_2 slowly.
- Safety matches are made by dipping the head of a match stick in potassium chlorate paste. The striking surface is made up of red phosphorus and sand.

NOBLE GASES

The elements

Helium	-	He (Z=2)
Neon	-	Ne (10)
Argon	-	Ar (18)
Krypton	-	Kr (36)
Xenon	-	Xe (54)
Radon	-	Rn (86) (radioactive)

Constitute group zero of the periodic table. All of them are monoatomic gases. These are called inert gases because of their chemical inertness.

a) Discovery of He:

He was first of the noble gases to be discovered. Janssen and Lockyer observed a bright yellow line in the solar spectrum which was attributed to He by Frankland & Lockyer.

b) Discovery of Ar:

Ar was discovered by Lord Rayleigh & Ramsay. A certain quantity of air, after the removal of moisture and CO_2 was passed repeatedly over red hot cooper to remove all O_2, and then over red hot Mg to remove all N_2. The residual gas which was nearly 1% of the volume of air taken was found to be completely un-reactive, Ramsay called the gas as argon.

c) Discovery of Ne, Kr & Xe:

Ramsay & Travers, carried out careful fractional evaporation of liquid Ar under varying reduced pressures and isolated three more gases i.e. Ne, Kr, Xe.

(i) Occurrence:

He is found as natural gas.

He and Ar are also found in the dissolved gases of some mineral springs.

Ar is the most abundant noble gas in the atmosphere.

Physical Properties:

1) Electronic configuration: nS^2nP^6

All the noble gases except He have eight electron in valence configuration.

2) Atomic radii:

Atomic radii of inert gases are highest in their respective periods. As we move down the group from Ne to Rn, the size of noble gases further increases primarily due to an increase in the number of shells.

The atomic radii increase from He to Xe:

$$He < Ne < Ar < Kr < Xe$$

3) Ionization energy:

IE of noble gases are highest in their respective periods due to their stable electronic configuration. IE decrease from He to Xe because of increase in atomic radii.

$$He > Ne > Ar > Kr > Xe > Rn$$

4) Electron affinity:

EA of noble gases is zero. This is because of the completely filled orbitals of all atoms of the noble gases which do not practically accept electron due to the extra stability associated with them.

5) M.P & B.P:

M.P & B.P of noble gases are quite low as compared to other elements of comparable atomic masses. M.P & B.P of these gases show a regular increase from

He to Rn due to an increase in magnitude of the Vander Waal's forces of attraction as the size of the atom increases.

$$He < Ne < Ar < Kr < Xe$$

6) Ease of liquification:

These gases cannot be easily liquefied at low temperature since their atoms are held together by weak Vander Waal's forces of attraction. The heats of vaporization increase from He to Rn.

$$He < Ne < Ar < Kr < Xe < Rn$$

7) Solubility:

These gases are sparingly soluble in water. The solubility increases from He to Xe.

$$He < Ne < Ar < Kr < Xe$$

8) Monoatomic nature:

Noble gases exist only in monoatomic state due to the non availability of unpaired of electron. The ratio of their specific heats at constant pressure (Cp) and constant volume (Cv) i.e Cp/Cv is 1.667.

9) Adsorption:

All the noble gases are adsorbed by activated wood charcoal at low temperature. Activated coconut-charcoal also adsorbs different gases at different temperature. Adsorbability increases with increase in the atomic mass.

$$He < Ne < Ar < Kr < Xe$$

10) Polarization:

The polarizability of the noble gases increases down the group.

$$He < Ne < Ar < Kr < Xe$$

11) Diffusion:

The ease of diffusion of noble gases decreases down the group as the size increase from He to Xe.

$$He > Ne > Ar > Kr > Xe$$

12) Electrical conductivity:

Noble gases have fairly high electrical conductivity. The gases produce characteristic, coloured lights. For ex: Neon gives brilliant orange red glow Hg and Ne gives a blue or green glow.

13) Thermal conductivity:

Thermal conductivities of noble gases at 273 K & atmospheric pressure decrease with increasing atomic size of the gas from He to Xe.

$$He > Ne > Ar > Kr > Xe$$

Chemical Properties:

Noble gases are chemically inert and usually do not form compounds because of their stable electronic configuration, high IE and almost zero EA. Thus they show zero valency.

1. Formation of Clathrates:

When noble gases get entrapped in the cavities of crystal lattices of certain organic and inorganic substances called "host" cage compounds (or) clathrates are formed. The only type of ineraction in these cage compounds is the relatively weak Vander Waals forces and there is no chemical bonding. Clathrate compounds of Ar, Kr and Xe have been isolated by locking up these gases in the cages of water, phenol (or) quinol. He and Ne do not form any clatharates.

2. Formation of interstitial compounds:

Compounds in which small atoms of non-metals occupy positions in the interstices of metal lattices, are called interstitial compounds. He due to its small size which matches the size of the interstices of most of the heavy metals has a very high tendency to form such compounds. Other noble gases do not form interstitial compounds due to their larger size.

3. Compounds of Xe:

a) Fluorides:

      Xe forms three compounds with fluorine. These are:

      Xenon difluorides (XeF_2)

      Xenon tetrafluoride (XeF_4)

      Xenon hexafluoride (XeF_6)

(i) Xe F_2:

      XeF_2 is formed when a mixture of Xe and F_2 in the ratio 1:3 by volume is passed through a nickel tube at 673 k.

$$Xe + F_2 \xrightarrow{Ni\ 673\ K} Xe\,F_2$$

Structure:

      Xe F_2 has trigonal bipyramid geometry due to Sp^3d hybridization of Xe. Three equatorial positions are occupied by lone pairs of electron giving a linear shape to the molecule.

Properties:

XeF$_2$ is a colourless crystalline solid, reacts with H$_2$ to give X$_3$ and HF. It is hydrolysed completely by H$_2$O.

$$2\ XeF_2 + 2H_2O \rightarrow 2\ Xe + O_2 + 4HF$$

It is act as mild fluorinating agent.

(ii) XeF$_4$:

XeF$_4$ is prepared by heating a mixture of Xe and F$_2$ in the ratio 1:5 in a nickel vessel at 673 K and then suddenly cooling it in acetone.

$$Xe + 2F_2 \xrightarrow{Ni,673\ K} XeF_4$$

Structure:

XeF$_4$ has square planar shape due to Sp^{3}d^2 hybridisation.

Properties:

XeF$_4$ is colourless, crystalline solid, soluble in anhydrous HF, reacts with H$_2$ to form Xe & HF and reacts with water to give highly explosive solid, XeO$_3$, partial hydrolysis yields XeOF$_2$.

$$XeF_4 + H_2O \xrightarrow{193\ K} XeOF_2 + 2HF$$

It acts as strong fluorinating agent.

(iii) XeF$_6$:

XeF$_6$ is prepared by heating a mixture of Xe and F$_2$ in the ratio 1:20 at 473-523 K under a pressure of 50 atm.

$$Xe + 3\ F_2 \xrightarrow[50\ atm]{473 - 823\ K} XeF_6$$

Structure:

XeF$_6$ has pentagonal bipyramid geometry due to Sp^{3}d^3 hybridisation.

Properties:

It is colourless, crystalline solid, highly soluble in anhydrous HF giving solution which is a good conductor of electricity.

$$HF + XeF_6 \rightarrow Xe F_5^+ + HF_2^-$$

Partial hydrolysis of XeF_6 yields $XeOF_4$ and complete hydrolysis yields XeO_3.

XeF_6 is the most powerful fluorinating agent.

(b) Oxides:

Xe forms two oxides such as xenon trioxide (XeO) and xenon tetraoxide (XeO_4).

$$6 XeF_4 + 12 H_2O \rightarrow 2 XeO_3 + 4 Xe + 3O_2 + 24 HF$$

$$XeF_6 + 3H_2O \rightarrow XeO_3 + 6HF$$

Structure:

XeO_3 has tetrahedral geometry due to Sp^3 hybridisation of Xe. Xe has a lone pair of electrons giving a trigonal pyramidal shape.

Properties:

It is a colourless solid, highly explosive and powerful oxidizing agent.

(ii) XeO_4:

XeO_4 is prepared by the action of concentration H_2SO_4 on sodium or barium xenate (Na_4XeO_6; Ba_2XeO_6) at room temperature.

$$Na_4XeO_6 + 2H_2SO_4 \rightarrow XeO_4 + 2Na_2SO_4 + 2H_2O$$

$$Ba_2XeO_6 + 2H_2SO_4 \rightarrow XeO_4 + 2BaSO_4 + 2H_2O$$

XeO_4 is purified by vacuum sublimation at 195 K.

Structure:

XeO$_4$ has tetrahedral structure due to Sp^3 hybridization.

Properties:

It is quite unstable gas and decomposes to Xe & O$_2$.

$$XeO_4 \rightarrow Xe + 2O_2$$

(c) Oxy fluorides:

Xe forms three types of fluorides such as

Xenon oxydifluoride (XeOF$_2$)

Xenon oxytetrafluoride (XeOF$_4$)

Xenon dioxydifluoride (XeO$_2$F$_2$)

(1) XeOF$_2$:

XeOF$_2$ is formed by partial hydrolysis of XeF$_4$ at 193 K.

$$XeF_4 + H_2O \xrightarrow{193\ K} XeOF_2 + 2HF$$

Structure:

XeOF$_2$ has trigonal bipyramid geometry due to Sp^{3}d hybridization of Xe. Two equatorial positions are occupied by lone pairs of electron giving T-shape to the molecule.

(2) XeOF$_4$:

XeOF$_4$ is prepared by partial hydrolysis of XeF$_6$.

$$XeF_6 + H_2O \rightarrow XeOF_4 + 2HF$$

It can also be prepared by the reaction of SiO$_2$ with XeF$_6$.

$$2\ XeF_6 + SiO_2 \rightarrow 2XeOF_4 + SiF_4$$

Structure:

XeOF$_4$ has octahedral geometry due to Sp^{3}d^2 hybridization of Xe.

Properties:

It is a colourless volatile liquid which melts at 227 K. It reacts with water to give XeO$_2$F$_2$ & XeO$_3$.

$$XeOF_4 + H_2O \rightarrow XeO_2F_2 + 2HF$$

$$XeO_2F_2 + H_2O \rightarrow XeO_3 + 2HF$$

(3) XeO$_2$F$_2$:

XeO$_2$F$_2$ is formed by partial hydrolysis of XeOF$_4$ (or) XeF$_6$.

$$XeOF_4 + H_2O \rightarrow XeO_2F_2 + 2HF$$

$$XeF_6 + 2H_2O \rightarrow XeO_2F_2 + 4HF$$

Structure:

XeO$_2$F$_2$ has trigonal bipyramid geometry due to Sp^{3}d hybridization of Xe.

Properties:

It is a colourless solid which melts at 303 K. It is easily hydrolysed to give XeO$_3$.

Uses of Noble gases:

1) Helium

He is the most widely used of all the noble gases. It is used for filling of ballons & air ships because of its non-inflammability and high lifting power.

Oxygen-helium (1:4) mixture is used for treatment of asthma and for artificial respiration in deep sea diving because unlike nitrogen, helium is not soluble in blood even under high pressure.

Neon:

Ne is widely used in neon-signs which are used for advertising purposes in the form of brilliant orange red glow. This colour is changed by mixing Ar & Hg vapours with Ne to light blue. Since the light of Ne signs has better penetrating power through fog & mist so these are used in beacon lights for safety of air navigations.

Argon:

Ar is used for creating inert atmosphere in chemical reactions, welding & metallurgical operations and for filling in incandescent and fluorescent lamps. It is also used in filling Geiger counter tubes and thermionic tubes.

Krypton & Xenon:

Kr & Xe are used in gas filled lamps, these gases are superior to Ar but are very costly. A mixture of Kr & Xe is also used in some flash tubes for high speed photography.

Radon:

Rn is used in radioactive research and therapeutics and in the non-surgical treatment of cancer and other maligner growths.

- XeO_3 explodes violently when dry and its explosion power is 22 times more than TNT.
- Clatharates play an important role in the separation of noble gases.
- The clatharates are convenient form of handling, processing and transporting of isotopes of noble gases.
- Ne lamps are used in botanical gardens and the green houses as it stimulates growth and is effective in the formation of chlorophyll.
- XeF_6 can't be stored in glass vessels because of the following reactions which finally give the dangerously explosive XeO_3.

$$2 \, XeF_6 + SiO_2 \rightarrow 2 \, XeOF_4 + SiF_4$$
$$2 \, XeOF_4 + SiO_2 \rightarrow 2 \, XeO_2F_2 + SiF_4$$
$$2 \, XeO_2F_2 + SiO_2 \rightarrow 2 \, XeO_3 + SiF_4$$

- Noble gases neither act as reducing agents nor as oxidizing agents.

TRANSITION ELEMENTS INCLUDING

LANTHANIDES AND ACTINIDES

Transition elements are the elements which lie between most electropositive s- and most electronmegative p-blocks in the long form of the periodic table. These are also called d-block elements because the last electron in them enters in the d-orbitals of the (n-1) or penultimate (last but one) shell.

A transition element may be defined as an element whose atom in the ground state or ion in one of the common O.S. has incomplete d-sub shell i.e. has electron between 1 to 9.

General electronic configuration: $(n-1)d^{1-10} nS^{0-2}$

The definition excludes Zn, Cd and Hg from transition elements because they do not have incomplete d-sub shell in the atomic state.

Some exceptional electronic configurations of transition elements are as follows:

i) $Cr = 3d^5 4S^1$,

   $Cu = 3d^{10} 4S^1$

ii) $Nb = 4d^4 5S^1$,

   $Mo = 4d^5 5S^1$,

   $Ru = 4d^7 5S^1$,

   $Rh = 4^8 5S^1$,

   $Pd = 4d^{10} 5S^0$,

   $Ag = 4d^{10} 5S^1$

iii) $Pt = 5d^9 6S^1$,

   $Au = 5d^{10} 6S^1$

The irregularities in the observed configurations of Cr, Cu, Mo, Pd, Ag & Au are explained on the basis of the completely and half filled orbitals which are more stable than other d-orbital configurations. This can be explained by other factors such as:

(i)   Nuclear-electron attraction

(ii)   Shielding of one electron by several other electrons

(iii) Inter electronic repulsion

(iv)   Exchange energy forces

General Characteristics:

Transition elements resemble in their physical and chemical properties as they have similar nS^2 orbital electronic configuration in the outer most energy shell. Thus they are hard metals and are good conductors.

Some of the general characteristics:

(1) Atomic and ionic radii:

The atomic radii of transition elements show the following trends.

(i)   They lie in between those 'S' and 'P' block elements.

(ii)   In a series, the atomic radii first decrease with increase in atomic number upto the middle of the series, then become constant & at the end of the series show a slight increase.   Atomic radius decreases as the nuclear charge increases.

3d Series	SC	Ti	V	Cr	Mn	Fe	Co	Ni	Cu	Zn
4d Series	Y	Zr	Nb	Mo	Tc	Ru	Rh	Pd	Ag	Cd
5d series	La	Hf	Ta	W	Re	Os	`Ir	Pt	Au	Hg

(iii) The atomic radius increase down the group. The atomic radii decreases from Ti to Cu due to increase in atomic number as shown below:

$$Ti^{+2} > V^{+2} > Cr^{+2} > Mn^{+2} > Fe^{+2} > Co^+ > Ni^{+2} > Cu^{+2} > Zn^{+2}$$

2) Metallic character:

All the transition elements are metals having hcp, ccp or bcc lattices. Metallic bonding is caused due to the presence of one or two electron in the outer most shell or energy level (nS) and also unpaired d– electron.

Eg: Cr, Mo & W -   Stronger metallic bonding & hard metals

        Zn, Cd & Hg -   Weak metallic bonding & soft metals

(3)   M.P & B.P:

These metals have very high M.P & B.P due to stronger metallic bonding. The M.P of the transition elements first rise to a maximum and then fall as the atomic number increases. Mn and Tc have abnormally low M.P. In a particular series, the metallic bond strength increases upto the middle with increasing number of unpaired electrons.

(4)   Density:

Because of large number of valence electron involving nS and (n-1)d orbitals in transition metals, they have strong metallic bonding due to which these metals possess high densities.

            Eg: Os   -   highest density

                    Sc   -   lowest density

(5)   Ionization energies:

First IE of d-block elements are higher than those of S-block elements and are lesser than those of p-block elements. IE increases from Sc to Zn.

            $SC < Ti < V < Cr < Mn < Fe < Co < Ni < Cu < Zn$

(6)   Thermodynamic stability:

Thermodynamic stability of the compounds depends upon the values of IE of the metals. Smaller the IE of the metals, stabler is its compound.

(7)   Electrode potentials & Reducing character:

69

Quantitatively the stability of transition metal ions in different O.S. (oxidation states) in solution can be determined on the basis of electrode potential data. The lower the electrode potential of the electrode, more stable is the oxidation state of the transition metal ion in aqueous solution. Electrode potential values depend upon energy of sublimation of the metal, the IE and the hydration energy.

All the elements of 3d series are good reducing agents except Cu. However they are weaker reducing agents than S-block elements.

(8)   Oxidation States (O.S.):

All the transition metals except the first and last member in each series show variable oxidation states:

(i)   The highest O.S. of transition metals are found in fluorides and oxides. The highest O.S. shown by any transition element is +8. Both Os & Ru show +8 O.S.

(ii)   The most common O.S. of 3d-transition metals is +2 which shows +3 O.S.

(iii) Mostly ionic bonds are formed in +2 and +3 O.S.

       Eg: MnO_4^- & CrO_4^{-2} where Mn & Cr show +7 & +6 oxidation state.

(iv)   Higher O.S. are stabilized by atoms of high E.N.

(v)   In going down a group, the stability of higher O.S. increases while that of lower O.S. decreases.

(vi)   Transition metals also show low O.S. of +1 in their compounds.

(9)   Catalytic Properties:

Many transition metals (like Co, Ni, Pt, Fe, Mo etc.) and their compounds are used as catalysts because of the following reasons:

(i)   Because of variable oxidation states, they easily absorb & reemit wide range of energies to provide the necessary activation energy.

(ii)   Because of variable O.S., they easily combine with one of the reactants to form an intermediate which reacts with the second reactant to form the final products.

V_2O_5 -   Manufacture of H_2SO_4 by contact process

Co   -   Used in the Fischer Tropsch process in the synthesis of gasoline (petrol)

Ni   -   Used in the hydrogenation of oils in to fats

Pt   -   Used in the manufacture of H_2SO_4 by contact process and in the oxidation of ammonia of nitric acid by Ostwald's process

Fe   -   Used in the synthesis of ammonia by Haber's process

Mo   -   Used as a promoter

10)   Coloured ions:

Most of the transition metal compounds are coloured both in the solid state & in aqueous solution.   This is because of the presence of incompletely filled d-orbitals.

No. of unpaired electrons	Ion	Colour
0	Sc^{+3}	Colourless
1	Ti^{+3}	Purple
2	V^{+3}	Green
3	Cr^{+3}	Violet
4	Mn^{+3}	Violet

5	Mn^{+2}	Pale pink
5	Fe^{+3}	Pale Violet
6	Fe^{+2}	Green
7	Co^{+2}	Pink
8	Ni^{+2}	Green
9	Cu^{+2}	Blue
10	Zn^{+2}	Colourless

Sc^{+3}, Ti^{+4} and V - have completely empty d-orbitals – Colourless

Cu^+, Ag^+, Au^+, Zn^{+2}, Cd^{+2} & Hg^{+2} – have completely filled d-orbitals – Colourless

11) Magnetic properties:

Due to the presence of unpaired electron in the (n-1) d-orbitals, the most of the transition metal ions & their compounds are paramagnetic, i.e. they are attracted by the magnetic field. As the number of unpaired electron increase from 1 to 5, the magnetic moment and hence paramagnetic character also increases. Those transition elements which have paired electrons are diamagnetic.

A paramagnetic substance is characterized by its effective magnetic moment (μ), using formula

$$\mu = \sqrt{n(n+2)} \ B.M.$$

Ion	Sc^{+3}, Ti^{+4}	Ti^{+3}	Ti^{+2}	V^{+2}	Mn^{+2}	Fe^{+2}
No. of unpaired electron	0	1	2	3	4	5
μ (B.M)	0	1.73	2.83	3.87	4.90	5.92

12) Complex formation:

Transition metal ions form a large number of complexes in which central metal ion is linked to a no. of ligands. This is because of the following characteristic properties:

(i)   They have high nuclear charge & small size.

(ii)   They have empty d-orbitals to accept the lone pairs of electron donated by the ligands.

The stability of complexes increases with increase in atomic number of the element in a series and with decreasing size. Moreover, higher valent cations form more stable complexes.

13)   Interstitial Compounds:

Transition metals form a number of interstitial compounds in which small non-metal atoms such as H, C, B, N and He occupy the empty spaces in their lattices and also forms bonds with them. Steel & Cast iron which are interstitial compounds of Fe & C are hard where malleability and ductility of the metal decreases.

14)   Alloy formation:

Due to similarity in atomic sizes, atoms of one transition metal can easily take up positions in the crystal lattice of other in the molten state and are miscible with each other forming solid solutions & smooth alloys on cooling. Alloys are generally harder, have higher melting points & more resistant to corrosion than the individual metals.

15)   Non-Stoichiometric compounds:

Non-stoichiometric compounds are those in which the chemical composition does not correspond to their ideal chemical formulae. The compounds of transition metals with O, S, Se and Te are generally non-stoichiometric.

7) General properties of first row transition metal compounds:

a) Oxides:

(i)   Oxides of metals in low O.S. +2 & +3 (Mo, M_3O_4) are generally basic except Cr_2O_3 which is amphoteric in character.

(ii)   Oxides of metals in higher oxidation states +5 are generally acidic in nature.

(iii)   Oxides of metals in their intermediate O.S. +4 are generally amphoteric in nature.

b) Halides:

(i)   The order of reactivity of halogens with 3d-transition metals is $F > Cl > Br > I$.

(ii)   Metals form fluorides in their highest oxidation states.

(iii)   Halides in higher O.S. of the metals get easily hydrolysed giving with fumes in air due to the formation of hydrogen halides.

$$TiCl_4 + 2H_2O \rightarrow TiO_2 + 4\,HCl$$

c) Sulphides:

(i)   Sulphides are formed with metals in their lower O.S. as sulphur is not a good oxidizing agent.

(ii)   Sulphides are usually coloured and are insoluble in water.

Eg: MnS is flesh coloured

CoS, NiS and CuS are black coloured

ZnS is dirty white in colour

8) Some compounds of transition elements:

(1)   Potassium dichromate ($K_2Cr_2O_7$)

Preparation:

It is prepared from chromite ore ($FeCr_2O_4$ (or) $FeOCr_2O_3$) through the following steps:

$$4\ FeCr_2O_4 + 16\ NaOH + 7O_2 \rightarrow 8Na_2CrO_4 + 2Fe_2O_3 + 8H_2O$$

$$2\ Na_2CrO_4 + H_2SO_4 \rightarrow Na_2Cr_2O_7 + Na_2SO_4 + H_2O$$

$$Na_2Cr_2O_7 + 2\ KCl \rightarrow K_2Cr_2O_7 + 2\ NaCl$$

Potassium dichromate being less soluble is obtained by fractional crystallization.

Properties:

(i)   These are orange red coloured crystals soluble in water.

(ii)   It is strong oxidizing agent in the acidic medium.

$$K_2Cr_2O_7 + 4\ H_2SO_4 \rightarrow K_2SO_4 + Cr_2(SO_4)_3 + 4H_2O + 3\ (O)$$

(iii) Action of heat:

When heated it decomposes with evolution of oxygen.

$$4\ K_2Cr_2O_7 + 2KOH \rightarrow 4\ K_2CrO_4 + 2\ Cr_2O_3 + 3O_2$$

(iv)   Reaction with H_2O_2:

Acidified solution of dichromate ion forms a deep blue colour with H_2O_2 due to the formation of peroxo compound $[CrO(O_2)_2]$ or CrO_5. The blue colour fades away due to the decomposition of CrO_5 to Cr^{+3} & Oxygen.

$$Cr_2O_7^{-2} + 4H_2O_2 + 2H^+ \rightarrow 2CrO_5 + 5H_2O$$

Uses:

Volumetric analysis, in chrome tanning in leather industry, in the preparation of chrome dialum, in printing and in organic chemistry as an oxidizing agent.

(2)   Potassium Permanganate ($KMnO_4$):

Preparation:

It is prepared by the pyrolusite ore (MnO_2) through the following step:

(i)   Conversion of pyrolusite into potassium manganate:

$$2\ MnO_2 + 4\ KOH + O_2 \rightarrow 2K_2MnO_4 + 2H_2O$$

$$2\ MnO_2 + 2K_2CO_3 + O_2 \rightarrow 2K_2MnO_4 + 2CO_2$$

$$MnO_2 + 2\ KOH + KNO_3 \rightarrow K_2MnO_4 + KNO_2 + H_2O$$

$$3\ MnO_2 + 6\ KOH + KClO_3 \rightarrow 3K_2MnO_4 + KCl + 2H_2O$$

(ii)   Oxidation of potassium manganate to potassium permanganate.

$$3\ K_2MnO_4 + 2\ CO_2 \rightarrow 2KMnO_4 + MnO_2 \uparrow + 2K_2CO_3$$

$$2\ KMnO_4 + H_2O + O_3 \rightarrow 2KMnO_4 + 2KOH + O_2$$

$$2\ KMnO_4 + Cl_2 \rightarrow 2KMnO_4 + 2KCl$$

The solution is filtered and evaporated to get deep purple black crystals of $KMnO_4$.

(i)   It is a deep purple crystalline solid, moderately soluble in water at room temperature.

(ii)   When heated to 746 K, it readily decomposes giving oxygen.

(iii)   With well cooled conc. H_2SO_4, $KMnO_4$ gives Mn_2O_7 (an explosive oil) which on warming decomposes to MnO_2.

$$2\ KMnO_4 + 2H_2SO_4 \rightarrow Mn_2O_7 + 2KHSO_4 + H_2O$$

(iv)   When heated in a current of H_2, solid $KMnO_4$ gives KOH, MnO and water vapours.

$$2 KMnO_4 + 5H_2 \rightarrow 2KOH + 2MnO + 4H_2O$$

(v)   Oxidising Property:

KMnO_4 is a powerful oxidizing agent.

a)   In neutral solution. It acts as a moderate oxidizing agent because of the reaction.

$$2 KMnO_4 + H_2O \rightarrow 2KOH + 2MnO_4 + 3O$$

(or) $MnO_4^- + 2H_2O + 3electron \rightarrow MnO_2 + 4OH^-$

b)   In alkaline solution, it act as a strong oxidizing agent because of the reaction.

$$2 KMnO_4 + 2KOH \rightarrow 2K_2MnO_4 + H_2O + O$$

(or) $MnO_4^- + electron \rightarrow MnO_4^{-2}$

K_2MnO_4 is further reduced to MnO_2 when a reducing agent is present.

$$K_2MnO_4 + H_2O \rightarrow MnO_2 + 2KOH + O$$

(or) $MnO_4^- + 2H_2O + 2electron \rightarrow MnO_2 + 4OH^-$

(c)   In acidic medium, in presence of dil. H_2SO_4, it acts as a strong oxidizing agent because of the reaction.

$$2KMnO_4 + 3H_2SO_4 \rightarrow K_2SO_4 + 2MnSO_4 + 3H_2O + 5(0)$$

(or) $MnO_4^- + 8 H^+ + 5electron \rightarrow Mn^{+2} + 4H_2O$

Structure:

Mn in MnO_4^- undergoes Sp^3 hybridisation and hence four oxygen atoms are arranged tetrahedrally around manganese.

Uses:

In volumetric analysis, as a strong oxidizing agent in the laboratory as well as in industry as Baiyer's reagent for testing unsaturation, as disinfectant and germicide.

9) Lanthanides and Actinides:

a) Lanthanides:

The elements with atomic number 58 to 71 are called lanthanides (or) lanthanones (or) rare earths. These are involve the filling of 4f-orbitals. Their general electronic configuration is

$$[Xe]\ 4f^{1-14}\ 5d^{0-1}\ 6S^2$$

Promethium (Pm), At no. 61 is the only synthetic radioactive lanthanide.

Properties:

(i)   These are highly dense metals and possess high M.P.

(ii)   They form alloys easily with other metals especially iron.

(iii)   Most stable O.S. of lanthanides is +3 and O.S. +2 & +4 also exist but they revert to +3.

(iv)   Lanthanide contraction:

The regular decrease in the size of lanthanide ions from La+3 to Lu+3 is known as lanthanide contration. It is due to greater effect of the increased nuclear charge than that of screening effect.

    (a)   It results in slight variation in their chemical properties which helps in their separation by ion exchange methods.

    (b)   Each element beyond lanthanum has same atomic radius as that of the element lying above it in the same group.

    (c)   The covalent character of hydroxides of lanthanides increases as the size decreases from La^{+3} to Lu^{+3}. Hence the basic strength decreases.

(d)   Tendency to form stable complexes from La^{+3} to Lu^{+3} increases as the size decreases in that order.

(e)   There is a slight increase in EN of the trivalent ions from La to Lu.

(f)   Since the radius of Yb^{+3} ion is comparable to the heavier lanthanides Tb, Dy, Ho and Er.

(v)   Colour:

Most of the trivalent lanthanide ions are coloured both in solid state and in aqueous solution.

(vi)   Manetic properties:

All lanthanide ions with the exception of Lu and Ce are paramagnetic because they contain magnetic unpaired electron in the 4f orbitals. The magnetic moment of lanthanides are calculated by taking into consideration spin as well as orbital contributions and a more complex formula.

$$\mu = \sqrt{4 S(S + 1) + L (L + 1)} \; B.M.$$

which involves the orbitals quantum number L and spin quantum number S.

(vii)   Complex formation:

Although the lanthanide ion have a high charge (+3), yet the size of their ions is very large yielding small charge to size ratio. i.e. low charge density. As a consequence, they have poor tendency to form complexes. They form complexes mainly with strong chelating agents.

Electronic Configuration of Lanthanides

Name	Symbol	Al. No.	Configuration
Cerium	Ce	58	[Xe] $4f^2$ $5d^0$ $6S^2$
Praseodymium	Pr	59	[Xe] $4f^3$ $5d^0$ $6S^2$
Neodymium	Nd	60	[Xe] $4f^4$ $5d^0$ $6S^2$
Promethium	Pm	61	[Xe] $4f^5$ $5d^0$ $6S^2$
Samarium	Sm	62	[Xe] $4f^6$ $5d^0$ $6S^2$
Europium	Eu	63	[Xe] $4f^7$ $5d^1$ $6S^2$
Gadolinium	Gd	64	[Xe] $4f^8$ $5d^0$ $6S^2$
Terbium	Tb	65	[Xe] $4f^9$ $5d^0$ $6S^2$
Dysprosium	Dy	66	[Xe] $4f^{10}$ $5d^0$ $6S^2$
Holmium	Ho	67	[Xe] $4f^{11}$ $5d^0$ $6S^2$
Erbium	Er	68	[Xe] $4f^{12}$ $5d^0$ $6S^2$
Thulium	Tm	69	[Xe] $4f^{13}$ $5d^0$ $6S^2$
Ytterbium	Yb	70	[Xe] $4f^{14}$ $5d^0$ $6S^2$
Lutetium	Lu	71	[Xe] $4f^{14}$ $5d^1$ $6S^2$

Electronic Configuration of actinides

Name	Symbol	Al. No.	Configuration
Thorium	Th	90	[Rn] $5f^0$ $6d^2$ $7S^2$
Protactinium	Pa	91	[Rn] $5f^2$ $6d^1$ $7S^2$
Uranium	U	92	[Rn] $5f^3$ $6d^1$ $7S^2$
Neptunium	Np	93	[Rn] $5f^4$ $6d^1$ $7S^2$
Plutonium	Pu	94	[Rn] $5f^5$ $6d^0$ $7S^2$
Americium	Am	95	[Rn] $5f^6$ $6d^0$ $7S^2$
Curium	Cm	96	[Rn] $5f^7$ $6d^1$ $7S^2$
Berkelium	Bk	97	[Rn] $5f^8$ $6d^1$ $7S^2$
Californium	Cf	98	[Rn] $5f^9$ $6d^0$ $7S^2$
Einsteinium	Es	99	[Rn] $5f^{10}$ $6d^0$ $7S^2$
Fermium	Fm	100	[Rn] $5f^{11}$ $6d^0$ $7S^2$
Mendelevium	Md	101	[Rn] $5f^{12}$ $6d^0$ $7S^2$
Nobelium	No	102	[Rn] $5f^{13}$ $6d^0$ $7S^2$
Lawrencium	Lr	103	[Rn] $5f^{14}$ $6d^1$ $7S^2$

b)   Actinides:

The elements with atomic number 90 to 103 i.e. Thorium to Lawrencium are called actinides (or) actinones. These elements involve the filling of 5f orbitals. Their general electronic configuration is

$$[Rn]\ 5f^{1-14}\ 6d^{0-1}\ 7S^2$$

They includes three naturally occurring elements Th, Pa & U and eleven transuranium elements which are produced artificially by nuclear reactions. They are synthetic or man made elements. All actinides are radioactive.

Both lanthanides and actinides are collectively called f-block elements because last electron in them enters into f-orbitals of the antepenultimate shell. The general electronic configuration of f-block elements is

$$(n-2) \, f^{1-4} \, (n-1) \, d^{0-1} \, nS^2$$

Properties

(i)   Oxidation State:

The dominant oxidation state of actinides is +3 which shows increasing stability for the heavier elements. Np shows +7 O.S. but this is oxidizing and is reduced to the most stable state +5. Pu shows up to +7 and Am up to +6 but the most stable state drops to Pu (+4) and Am in +4 state due to f^7 configuration.

When the oxidation number increases to +6, the actinide ions are no longer simple. The exhibition of large number of O.S. of actinides is due to the fact that there is a very small energy gap between 5f, 6d & 7s subshells.

(ii)   Actinide contraction:

There is a regular decrease in ionic radii with increase in atomic number from Th to Lr. This is called actinide contraction analogous to the lanthanide contraction. It is caused due to imperfect shielding of one 5f electron by another in the small shell.   This results in increase in the effective nuclear charge which causes contraction in size of the electron cloud.

(iii)   **Colour of the ions:** Ions of actinides are generally coloured which is due to f-f transitions. It depends upon the number of electron in 5f orbitals.

(iv)   Magnetic Properties:

Like lanthanides, actinide elements are strongly paramagnetic. The magnetic moments are lesser than the theoretical predicted values. This is due to fact that 5f electron in actinides are less effectively shielded which results in quenching of orbital contribution.

(v)   Complex formation:

Actinides have a greater tendency to form complexes because of the higher nuclear charge and smaller size of their atoms. They form complexes even with π-bonding ligands such as alkyl phosphines, thioethers etc.

$$Mn^{+4} > MO_2^{+2} > M^{+3} > MO_2^{+}$$

The absorption of H_2 by transition metals such as Pt, Pd, Ni etc. is called occlusion and is due to the interstitial hydride formation.

Metal ion	Colour	Ion	Colour
La^{+3}	Colourless	Lu^{+3}	Colourless
Ce^{+3}	Colourless	Yb^{+3}	Colourless
Pr^{+3}	Yellow green	Tm^{+3}	Green
Nd^{+3}	Red	Er^{+3}	Pink
Pm^{+3}	Uncertain	Ho^{+3}	Yellow
Sm^{+3}	Yellow	Dy^{+3}	Yellow
Eu^{+3}	Pink	Tb^{+3}	Pink
		Gd^{+3}	Colourless

www.ingramcontent.com/pod-product-compliance
Lightning Source LLC
Chambersburg PA
CBHW040829180526
45159CB00001B/123